AND THE EARTH SHALL REEL TO AND FRO

THE PROPHECY TRILOGY - VOLUME II

ANTHONY E. LARSON

ZEDEK BOOKS
P.O. Box 1271 Orem, Utah 84057

First Printing, August 1983
Second Printing, May 1987
Third Printing, December 1989
Fourth Printing, May 1993

Zedek Books
Orem, Utah

Printed in the United States of America

Contents

Preface

And The Earth Shall Reel To And Fro has been published because the first book in this series, which introduced these novel interpretations of history and prophecy, was so well received by Latter-day Saints. Prior to the publication of *And The Moon Shall Turn To Blood* (the first volume of the Prophecy Trilogy series), some critics claimed that the topic would be too controversial, too speculative to be well received. Others said that the material was too weighty or too scholarly for the average Latter-day Saint. Thankfully, the critics have been proven wrong. Thousands of Latter-day Saints have read *And The Moon Shall Turn To Blood,* and a great many have expressed their approval and enthusiasm for this new insight into prophecy and history. In fact, the overwhelming majority of the responses I have received were in the form of requests for more information on the subject.

And The Earth Shall Reel To And Fro was written in response to the desire among Latter-day Saints for more information dealing with catastrophic events in history and prophecy. It is intended to be a companion volume to the book *And The Moon Shall Turn To Blood* and is a continuation of the concepts examined in that initial volume. Indeed, this book may be considered the second volume in a series dealing with history and prophecy from a catastrophist viewpoint.

Present plans include yet a third book in this series, making this set a trilogy on prophecy. It will deal with the coming, radical changes in our solar system and how they will affect the future condition of this planet and its inhabitants.

For the uninitiated reader to become acquainted with all the concepts of this book, it is suggested that the reader first read *And The Moon Shall Turn To Blood.* As with any series, later volumes build upon the fundamental precepts set forth in the preceding volumes. Without reading the introductory material found in the first book, the uninitiated reader might find the material presented in this book to be speculative and without substance or foundation. Yet, with the proper introduction provided by the first book in this series, this thesis not only appears logical and rational, but will serve to solve some of the most puzzling questions regarding history and prophecy.

And they shall see signs and wonders, for they shall be shown forth in the heavens above, and in the earth beneath.

Doctrine & Covenants 45:40

1

An Introduction

The Lord has said: ". . . there is no God beside me, and all things are present with me, for I know them all." (Moses 1:6) Further, Joseph Smith wrote: ". . . the past, the present, and the future were and are, with Him, one eternal 'now' . . . ." (*Teachings of the Prophet Joseph Smith,* p. 220). Those two statements may serve to explain why the Lord uses similar terms to describe both past and future events. It may be, in fact, that "there is nothing new under the sun," that the future is nothing but a repetition of past events.

Upon examining the language used to describe past catastrophes (such as the Exodus, the Flood, the long day of Joshua, or the destruction of Sodom and Gomorrah), and comparing it with the language used to describe future catastrophes (like those described by John in Revelation, Isaiah, Zechariah, or Joseph Smith in the Doctrine & Covenants), it becomes apparent that the Lord describes both past and future events using identical similes and metaphors. This similarity, then, becomes a key to prophecy. If we can grasp the true nature and extent of past catastrophes, then we have begun mastering an understanding of future catastrophes, because they are described in identical terms.

To that end, in this volume (as in the first), both past and future catastrophes will be analyzed. Several books from scripture will be examined individually in order to emphasize the continuity of the concepts as they were presented by the prophets themselves. This should plainly demonstrate that these ideas and concepts have not been wrested from the scriptures (by quoting out of context or by inferring something other than what the prophets intended to say). It should become evident to the reader that nothing is being read into the scriptures; but rather, that the scriptures themselves are excellent witnesses to the evidence of great ancient catastrophes which reposes within them. Though space will not permit a thorough examination of all the writings in the scriptures, from a catastrophist point of view, those books selected for examination are typical of the remainder.

Some chapters in this book deal solely with the writings of individual prophets, in order to grasp the message they wished to communicate. By examining those writings individually, the flavor and intent of each prophet's message comes through. At the same time, other chapters synthesize the statements of many prophets on a particular aspect of future events, in order to present a composite picture of what coming catastrophes will be like, and how to prepare for them.

A word of caution

The reader would do well to remember that ancient and modern prophets wrote of catastrophic events in their own times, in times before them, and in times to come. Indeed, in their visions, they were given the "now" perspective of the past and the future that the Lord has; hence, they described their experiences from this same perspective. Because this book deals with those catastrophic events from the prophets' perspective, the reader may find himself reading of both past and future events in the same page—sometimes in the same paragraph. The prophets habitually mixed their references to the past and the future. Though this may prove somewhat confusing at first, such treatment of the material is necessary in order to see things "through the eyes of the prophets," remembering that they undoubtedly were shown the past and the future from the same perspective the Lord sees them—from the "now" perspective. Once the reader understands that same perspective, he can easily grasp the prophetic message of the prophets and decipher the same references in any other book of scripture, ancient or modern.

The topics are wide-ranging, diverse, and touch on a multitude of disciplines. The reader should not be dismayed in being confronted with archaeology one minute and prophecy the next, astronomy on the one hand and mythology on the other. Every effort has been made to present the material in a plain and orderly manner so as to be comprehensible to the layman. Indeed, the author himself is but a layman, seeking to solve the puzzles that confound us. The emphasis here is on making the material as understandable as possible. Hopefully, this interpretation of history, prophecy, and science will make them all more intelligible and logical.

Learn what is to come

Part of the truth that we must deal with is the knowledge that the prophesied catastrophes will profoundly change every aspect of our lives and the world we live in. Hence, we should become familiar with the changes that past calamities wrought on the world so that we might know what to expect and how to deal with them when such things occur in our time.

The careful, detailed analyses of past catastrophes have provided a format—a blueprint, if you will—of how such events occur. Although each catastrophe had its own peculiarities, they all had several effects or manifestations in common, and they all shared this one commonality: every major, world-wide calamity was caused by extra-terrestrial agents—bodies in space. That common denominator provides the key to understanding the same type of calamities yet to come. Hence, because we have a rather accurate blueprint of past events, we also have an accurate blueprint of future events. And just as we might examine the plans for a building, to learn how it is to be assembled and what it will look like, we may examine the catastrophe model by examining past calamities and then compare our model to prophecy, in order to learn what the future holds for us.

We share the vision

The seers' vision—the ability to look backward and forward through the corridors of time—is a marvelous and unique gift given to only a few. It is true that no man can know the past or the future except by revelation—a gift from God. Those of us who have not yet received or developed this gift must depend upon those who have. Happily, we have the writings of many men—seers—who have seen the past and the future. And now, in addition, we also have a key to assist us in understanding what they saw and described. Their vision becomes our vision, too!

"Who is wise, and he shall understand these things? prudent, and he shall know them?"

Hosea 13:9

2

Habakkuk—A Song of Catastrophe

Habakkuk is one of the more obscure books in the Old Testament. The third chapter, specifically, contains an excellent description of the judgments of God seen to occur in past epochs. At the same time, it strongly resembles the prophetic descriptions of future events found elsewhere in the scriptures.

This third chapter was probably a poem—meant to be spoken or sung to music (which seems fitting since it is a dramatic presentation of spectacular events once witnessed by all mankind). Notice the last sentence: "To the chief singer on my stringed instruments." It appears to be an annotation directing the poem to a musician. In addition, scattered throughout the text, at the end of several paragraphs, the word "Selah" is found, meaning to pause or rest (much like rest marks in modern musical notation). It may have been a word used to divide the lyrics into verses. Anciently, this type of direction was written right in the text to indicate to the reader or singer how the writer wished the piece performed. (See *Smith's Bible Dictionary* by William Smith, under "Selah.")

This bit of information makes chapter 3 all the more interesting. It was meant to be sung or read to music. It was probably performed for the children of Israel in order to remind them of the magnificent deeds God had done in their behalf. Elsewhere in the Bible is recorded a very similar song which was sung by Moses and the children of Israel once they were safely on the opposite shore of the Red Sea. They praised God for their deliverance, singing, "Who is like unto thee, O Lord, among the gods? who is like thee, glorious in holiness, fearful in praises, doing wonders?" (Exodus 15:11.) Such poems or songs were commonly used anciently in many cultures to dramatize the events they depicted—usually heroic acts of men and gods. Hence, the acts of God in destroying the wicked, while safely delivering the Israelites, were memorialized in poems and songs like that found in Habakkuk. The Psalms are replete with such descriptions. The supernatural acts of God were ideal topics for songs. Properly performed, such a story would have been more dramatic than any other to the Israelites.

Both the song of Moses and the third chapter of Habakkuk tell of the might of God in the cataclysms and destructions of the past when worlds nearly collided. Habakkuk, however, chronicled many of the manifestations omitted from the Exodus account, which will be helpful when attempting to decipher history or prophecy.

A review of the past

Habakkuk began: "O Lord, revive thy work in the midst of the years, in the midst of the years make known; in wrath remember mercy." (Habakkuk 3:2.) The "work" of God to which Habakkuk referred is certainly the destruction which He poured out on the Earth from time to time. Subsequent verses will clearly demonstrate that these marvelous phenomena were attributed to the "wrath" and "anger" of God. (Habakkuk 3:8.)

Habakkuk asked the Lord to "revive" His work, that is, to repeat again what He had done before. That statement shows an understanding on the part of the prophet that these things do recur from time to time. Also, it may be speculative but noteworthy to indicate that Habakkuk did not ask the Lord to do these things in "the last days." Instead, he asked that these things be done again "in the midst of the years." The phrase "midst of the years" seems to refer to a period of time other than the last days, leaving the impression that we are dealing with history in this chapter—not prophecy. However, it bears repeating that the events depicted in this chapter of Habakkuk are indicative of cataclysms seen to occur in all ages; that is, these destructions, the "wrath of God," are strikingly similar whether we speak of history or prophecy.

In addition, Habakkuk asked that the Lord be merciful to them while pouring out destruction: "in wrath remember mercy." It is logical that the prophet would plead for mercy, considering the magnitude of the destructions he proceeded to describe.

Verse 3 begins: "God came from Teman, and the Holy One from mount Paran." Of course, God does not dwell in these two places (both south and southwest of Jerusalem), so it seems likely that the object seen, which caused these destructions, first appeared in the heavens in the direction of Teman and mount Paran. It is frequently the case that the scriptures refer to the position on the horizon where these manifestations were seen. (See Isaiah 14:12-14, Zechariah 14:4.)

Who or what was it?

The prophet here refers to the agent of these destructions as God, Himself. It is evident that though God is the prime mover in the Universe, no one actually saw God above the Earth in the heavens doing these things. What was seen was an agent of God—one of His creations. Understanding this manner of describing the object in heaven is fundamental to comprehending prophetic accounts. Often the prophets refer to these objects as God when they obviously knew that such was not the case. (This may have been simply a manner of speech. Today, in much the same manner, when a miraculous event occurs, we say, "I see the hand of the Lord in this," when no one has actually seen the hand of the Lord at all.)

Anthropomorphism—assigning human qualities, features, and attributes to inanimate objects—is a common ancient device. The gods of the Romans and the Greeks, for example, were all spoken of and worshipped as beings who looked and acted like men. Yet if Velikovsky's theories are correct, those gods were originally astral bodies engaged in conflict in the heavens. Later, when written records of the movements and conflicts of the planets/gods were made, mankind had embellished their exploits, clothing them with human bodies and endowing them with human emotions and weaknesses.

Since they saw the power of God demonstrated in these objects, the prophets referred to them variously as "God," "the Holy One," "the arm of the Lord," "the Angel of the Lord," etc. Hence, a reference to an "angel" or to "God" may mean *an actual being,* or it may mean *an inanimate agent.* Therefore, care must be exercised when reading these passages (and others like them), so as to not misunderstand whether the writer was referring to a being (God or His messengers), or an inanimate agent (a heavenly body). This determination can usually be made by simply referring to the context itself. In this case, the prophet is referring to a past encounter between Earth and another planet/comet-like body. This becomes further evident as the prophet Habakkuk lists the manifestations that are clearly connected to a planetary close encounter.

An awesome apparition

The next two sentences give a description of the comet/planet as it appeared in heaven:

> His glory covered the heavens, and the earth was full of his praise. And his brightness was as the light; he had horns coming out of his hand: and there was the hiding of his power. (Habakkuk 3:3-4.)

Velikovsky indicated that the protoplanet Venus—the "destroying angel" of Exodus—had an awesome appearance as it traversed the heavens. (See *And The Moon Shall Turn To Blood* for information on Dr. Velikovsky and his theories.) It had been recently born of Jupiter in what must have been an enormous cataclysm. It certainly had a fiery aspect, glowing brightly. In addition, the elements of its atmosphere (which formed its tremendous tail anciently), were highly reflective—sending back into space nearly all the sunlight that fell upon it. Even today, the pacific planet Venus reflects back into space most of the sunlight striking its atmosphere, making it one of the most luminous bodies in space. (As a measure of Venus's reflectivity, pristine snow—as bright as it is in direct sunlight—reflects less sunlight than does the atmosphere of Venus.) The material which currently composes the highly reflective atmosphere of Venus once stretched across space to form its tremendous tail. When the errant planet settled into its present orbit, the tail collapsed around the planet, forming the present atmosphere.

Anciently, the reflectivity of its giant tail and the fiery glow of the body or head, combined to produce an extraordinary sight. Isaiah described its brightness as exceeding that of the stars of heaven and even rivaling that of the Sun: "the most High," in splendor. (Isaiah 14:13-14.) Habakkuk also notes that, "His glory covered the heavens." Because this body was such a glorious and impressive manifestation, nations the world over began worshipping it—hence the statement by Habakkuk: "the earth was full of his praise."

It is most significant that the prophet declares that "he had horns coming out of his hand," because the idolatry instigated among Earth's inhabitants by the appearance of the comet/planet was characterized by worship of a deity with horns.

Velikovsky theorized that this was a natural consequence, due to the appearance of the comet of the Exodus. It had appeared in heaven with "horns." Our moon, for example, seen from the Earth, appears to go through phases. During the crescent phase, the moon appears with two pointed tips or "horns." Likewise, the great body that menaced the Earth anciently and rivaled the Sun in brightness, would have shown itself to Earth's inhabitants in various crescent phases—with "horns." But in addition, because this comet/planet had a tremendous tail, the tail would have served to greatly extend the "horns" of its crescent phase. When the Earth and the comet/planet were in proper relation to one another in space, Earth's inhabitants were treated to a celestial "sign" of imposing magnitude—the head of the celestial cow, calf, ox, or bull.

A pagan god

The apis cult (featuring the worship of a bovine god), was widespread anciently—instigated by the appearance of a comet/planet in the heavens. Without this insight, the statement by the prophet about horns makes no sense. Why would a prophet make reference to God showing the symbol of a pagan deity? Habakkuk was no pagan—he taught of Jehovah, the God of Israel. Why, then, would he use a pagan symbol with reference to the God of Israel? Simply because he wished to use the symbol familiar to all people.

The "heathen nations" (as the Hebrew prophets called them), commenced building and worshipping idols resembling the various aspects of the "god" they had seen in heaven. And although the Israelites were taught not to worship the "gods of the heathens," idolatry frequently surfaced among them, and the prophets constantly spoke out against it. Surrounded by nations which worshipped the "hosts of heaven" (the planetary gods), and having those among them who worshipped idols, the Jewish prophets used references which were familiar to all people anciently.

Hence, the prophet's reference to "horns coming out of his hand" was not a slip of the pen, nor was it a reference to idolatry on *his own part,* but it was a factual statement that all men in his day would understand concerning the appearance of the heavenly object both before and after the time of the Exodus. He used a universal symbol—an accepted heavenly sign.

Sure signs

Next, Habakkuk mentioned the now familiar signs of destruction:

> Before him went the pestilence, and burning coals went forth at his feet. He stood, and measured the earth: he beheld, and drove asunder the nations; and the everlasting mountains were scattered, the perpetual hills did bow: his ways are everlasting. (Habakkuk 3:5-6.)

First he mentioned the pestilence. This is nearly always one of the earliest signs—the very onset of destruction. Not only were the Exodus plagues of vermin characterized as a pestilence, but so was the plague of bloody water. And because the plagues of vermin and bloody water were seen to precede the plague of falling meteorites, the prophet next spoke of "burning coals." Other scriptural accounts tell of "hail" or "fire mingled with hail" (Exodus 9:24), or of "great stones" or "hailstones" (Joshua 10:11) falling from the sky, but Habakkuk used a more descriptive reference to the hot meteorites that pelted the Earth at the appearance of this comet. He called them "burning coals."

Moving mountains

Next, in proper sequence, came the most devastating sign—the unparalleled geologic activity of past celestial encounters, which the prophet described as having "drove asunder the nations" and "scattered" the "everlasting hills," so that "the perpetual hills did bow"—a most poetic description of what must have been a devastating sight to behold. Again, this is a universal sign always experienced by mankind as part of a tragically familiar series of phenomena.

The heavenly pillar

Another point is the prophet's reference to God as having feet and standing. Again, this was not God Himself. It was, rather, the appearance of one of His agents. The true nature of God's physical appearance has been revealed to Latter-day Saints. While He does have a physical body with feet and legs for standing, He is no giant being. The statement, "He stood, and measured the earth," evokes the image of a colossal being, standing (spread-legged) on the horizon. The imagery seems appropriate considering that the comet and its tail formed a perpendicular column, (wider at its base than at its peak) which appeared to stand above the Earth on the horizon. Moses described the celestial agent of the Exodus as "a pillar of a cloud" and declared that the Lord was in it (Exodus 13:21), and that it was "the angel of God." (Exodus 14:19.) It appears that Habakkuk's reference to the object having feet and standing to "measure the earth," together with Moses's reference to a "pillar," is an allusion to the appearance of the head and tail of the comet as it appeared to "stand" on the horizon like "a pillar of a cloud."

In verse 7, Habakkuk mentions that other areas bordering the land of Israel also suffered. "I saw the tents of Cushan in affliction: and the curtains of the land of Midian did tremble." (Habakkuk 3:7.) Such statements serve to underscore the global dimensions of the destruction the prophet described.

The effect on bodies of water

Velikovsky indicated that during past destructions, rivers, lakes, and oceans were adversely affected by the tidal action of the passing body. That phenomenon serves to explain the prophets rhetorical questions in verse 8:

> Was the Lord displeased against the rivers? was thine anger against the rivers? was thy wrath against the sea? (Habakkuk 3:8.)

Some rivers rose up out of their banks and took new courses; others even reversed their direction of flow; and still others ceased to flow altogether, according to Velikovsky's interpretation. Of course, the Old Testament graphically describes the recession and regression of the Red Sea at the time of the Exodus. (Exodus 14:19-31.)

Habakkuk's later statement makes it apparent that geologic changes were so great that new rivers were born:

> The mountains saw thee, and they trembled: the overflowing of the water passed by: the deep uttered his voice, and lifted up his hands on high. (Habakkuk 3:10.)

The reference to trembling mountains, together with the allusion to the unusual movement of bodies of water, is additional evidence of a cataclysmic event—cosmic in nature.

Note the three things that water was seen to do. An immense tide caused by the gravitational pull of the passing intruder would (1) overflow the land, (2) make a terrible roaring noise, and (3) stand high above normal sea level—appearing as a mountain of water standing out of the Earth. It must have been an awesome sight! In verse 15 the prophet calls it "the heap of great waters." (Habakkuk 3:15.) This is almost identical to the statement of Moses wherein he says, "The floods stood upright as an heap." (Exodus 15:8.)

The changeling comet

Then follows another reference to the appearance of the comet/planet: "Thy bow was made quite naked. . . ." (Habakkuk 3:9.) The crescent phase of the comet must also have appeared as a bow in heaven. Obviously, the Hebrew prophets who warned against the pagan worship of the ox or cow would have been anxious to designate the heavenly apparition as something other than the head of a bovine animal. It may have been inappropriate for the prophet to write, "Thy horns were made quite naked. . . ." Hence, the prophet chose the image of a "bow" in heaven to supplant the obnoxious imagery used by the idolaters.

If it seems to the reader that this object in heaven is made to appear in many forms—almost too many to be believable—it may be remembered that this object almost certainly moved on an elongated elliptical course through the heavens for hundreds of years before and after the Exodus. Because it traveled on an orbit very different from that of any other planet, it appeared to come and go in the heavens independent of the other stars or planets—much as ordinary comets do today. It also changed its appearance from time to time because its position in space, with relation to the Earth, would have been constantly changing. So it is easily conceivable that this errant comet/planet, after many encounters with other planets over the centuries, would have appeared to represent different earthly objects.

The appearance of comets

Today, ordinary comets, which follow orbits somewhat similar to that which the comet of the Exodus followed, are often seen to change appearance. Over the last 300 years myriad comets have appeared in the heavens, and records have been kept of their appearance. No two of them have looked exactly the same. Some had long, trailing tails. Some looked like a comma on a printed page. Others had split, divided, or parted tails. Some even looked like large crescents. The comet of 1744 had a six-streamered tail. Comet Sawerthal of 1888 had (in addition to a long, trailing tail) two other comma-shaped tails—each perpendicular to the main tail so that the comet appeared to have horns. The comet of 1528 was described by French surgeon Ambroise Pare:

> This comet was so horrible and so frightful and it produced such great terror in the vulgar that some died of fear and others fell sick. At the summit of it was seen the figure of a bent arm, holding in its hand a great sword, as if about to strike. On both sides of the rays of this comet were seen a great number of axes, knives and blood coloured swords among which were a great number of hideous human faces with beards and bristling hair. (As

cited by P. L. Brown in *Comets, Meteorites & Men,* pp. 10-14.)

The appearance of the comet/planet of the Exodus would have been even more dramatic when it came in close contact with the Earth and with other planets within the inner solar system. Earthbound skywatchers would have seen it go through many changes over the years, so that at various times it would have appeared as a horned god, a serpent, a dragon, a bow, a celestial crown, a sphere of light, an arm, a sword, etc. The descriptions ancients used to describe the many aspects of that heavenly apparition are multitude, but they are consistently found in the legends and myths of ancient cultures from all over the world.

Celestial sideshow

Then: "The sun and moon stood still in their habitation." (Habakkuk 3:11.) Earth's rotation stopped. This may seem to be a reference to the event in the days of Joshua, but Velikovsky maintained that it was also a manifestation seen to occur in the days of the Exodus. Since the same celestial body caused the events of the Exodus (and later, the events in the days of Joshua), it seems logical that the manifestations would be similar if not equally destructive. If the Earth stopped turning for a few hours in the days of Joshua, then logic dictates that it also stopped in the days of Moses, since (according to Velikovsky) the same passing body was responsible for both events. Whether the prophet Habakkuk was referring to the Exodus events or the Joshua events is unclear. Perhaps he was describing a synthesis of both events. Either way, this interpretation clearly serves to amplify our understanding of the nature of those cataclysms.

Tremendous electrical discharges in heaven also gave visible light. "At the light of thine arrows they went, and the shining of thy glittering spear." (Habakkuk 3:11.) Darkness covered the Earth to the degree that most of the light by which the survivors could see came from the luminous comet/planet. The form of the tail was compared to a "glittering spear." The bolts of lightning were characterized by Habakkuk as the "arrows" of God. All mythological history speaks of the arrows or thunderbolts of the gods. Such legends find their origins in the celestial firework displays of past planetary catastrophes, when huge bolts of interplanetary lightning lept between passing bodies.

Survivor's point of view

Verses 12 and 13 show the somewhat strange, but typical, feelings of the Hebrew prophets. They consistently taught that such destructions were sent to "thresh the heathen." At first glance it would appear that because these cataclysms were world-wide and universally destructive, the Israelites must have suffered from them as much as any other "heathen" people or nation. Nevertheless, the writings of the prophets consistently describe those destructions as judgements upon a particularly wicked segment of Earth's populace. Notice the prophet's feelings.

> Thou didst march through the land in indignation, thou didst thresh the heathen in anger. Thou wentest forth for the salvation of thy people, even for salvation with thine anointed; thou woundedst the head out of the house of the wicked, by discovering the foundation unto the neck. (Habakkuk 3:12-13.)

The Israelites considered those destructions to be an act of God in their behalf. Indeed, they had been shown great mercy, while other nations had suffered fully the wrath of the Almighty. Consider, for example, how the Egyptians were destroyed by the returning water of the Red Sea, while the Israelites had passed over, seemingly without mishap.

Although many individuals among the Israelites undoubtedly suffered and died because of these great natural calamities, it was the belief of the survivors that God was collectively "saving" the House of Israel. Many manifestations, such as the example of the parting of the Red Sea just cited, or the life-giving manna, served to reinforce that concept.

It is likely that their point of view sprang from the fact that the Israelites were forewarned of these events by the prophets and instructed by them in the measures to be taken in order to avoid injury or death. Those who were wise enough or faithful enough to follow the counsel of the prophets were the most likely to survive. Hence, the survivors saw themselves as the righteous remnant, while the wicked who failed to heed the prophet's warnings perished. Naturally, the survivors would write of themselves as having enjoyed the "mercy of God"—they had survived!

More apparitions

Another phenomenon connected with past celestial cataclysms, according to Velikovsky, was the appearance of smaller satellites seen to accompany the larger comet/planet. These satellites or small moons had trains or tails much like the contrails created by modern, high-flying jet aircraft. Therefore they appeared as though they were pulling the larger body along—much as a horse tethered to a buggy or chariot. Hence the prophet (in verse 15) makes such an analogy. (See chapters 3 and 4 on Isaiah and Joel for more on this phenomenon.)

Speaking of the passing body the prophet wrote: "Thou didst walk through the sea with thine horses, through the heap of great waters." (Habakkuk 3:15.) Another, earlier reference in this same chapter reads: "Was thy wrath against the sea, that thou didst ride upon thine horses and thy chariots of salvation?" (Habakkuk 3:8.) Similar references to celestial horses and chariots can be found in the myths and legends of cultures the world over, and they always appear in connection with other cataclysmic signs or events.

The prophet's fear

Verse 16 contains an observation of how one might feel when confronted with such tremendous natural events. It also contains a wish which likely we all share with the prophet:

> When I heard, my belly trembled; my lips quivered at the voice: rottenness entered into my bones, and I trembled in myself, that I might rest in the day of trouble. (Habakkuk 3:16.)

Rather than saying, "rottenness entered into my bones," we would probably say, "my knees grew weak."

The prophet's observation is a valid and graphic one. He saw and described his own emotional reaction to a cataclysmic event. At the same time, his reaction was typical for anyone experiencing these calamities firsthand. Living as we do at the end of a long period of peace in the solar system, it is difficult for us to comprehend these things happening to us. However, the prophet understood them very well and apparently felt distress at the thought of suffering the effects of a cosmic catastrophe. As we explore the words of the prophets, we get a more accurate picture of what it was like to experience these things firsthand.

At the end of verse 16, Habakkuk adds a warning to future generations: "When he cometh up unto the people, he will invade them with his troops." This is no earthly army. These are the hosts of heaven—the planets and other smaller bodies seen to accompany them. In other books of the scriptures, the prophets often tell of imposing celestial armies. (See Isaiah 5:26-30 and Joel 2:1-11, for example.) The army of God is a celestial army, not an earthly one. The "troops" of the heavenly army are bodies—large and small—from space. (Again, the reader is referred to chapters 3 and 4 for a more complete explanation of this phenomenon.)

Destroyed food

The last three verses of Habakkuk allude to the lack of foodstuff created by these cataclysms:

> Although the fig tree shall not blossom neither shall fruit be in the vines; the labour of the olive shall fail, and the fields shall yield no meat; the flock shall be cut off from the fold, and there shall be no herd in the stalls: Yet I will rejoice in the Lord, I will joy in the God of my salvation. (Habakkuk 3:17-18.)

Velikovsky indicated that little vegetation survived the destructions of the Exodus. Apparently, the darkness that shrouded the Earth during the Exodus, dissipated very slowly (over a period of many years). During that time, the vegetation of Earth languished for lack of direct sunlight. Habakkuk also tells of a time when the plants and animals that man depended on for food were either destroyed or refused to grow.

Remember that the survivors of the events of the Exodus were fed with manna—itself a byproduct of the planetary encounter. The survivors credited God (and properly so since He is the Author of these events) for the alimentation they needed when all other food sources were destroyed.

A comparison

A most interesting parallel to the book of Habakkuk (and equally obscure) is the book of Nahum. Compare the following verses from Nahum to those of Habakkuk:

> The Lord is slow to anger, and great in power, and will not at all acquit the wicked; the Lord hath his way in the whirlwind and in the storm, and the clouds are the dust of his feet.
>
> He rebuketh the sea, and maketh it dry, and drieth up all the rivers; Bashan languisheth, and Carmel, and the flower of Lebanon languisheth.
>
> The mountains quake at him, and the hills melt, and the earth is burned at his presence, yea, the world, and all that dwell therein.
>
> Who can stand before his indignation? and who can abide in the fierceness of his anger? his fury is poured out like fire, and the rocks are thrown down by him.
>
> The Lord is god, a strong hold in the day of trouble; and he knoweth them that trust in him.
>
> But with an overrunning flood he will make an utter end of the place thereof, and darkness shall pursue his enemies. (Nahum 1:3-8.)

The parallelism is plain and unmistakeable, yet Habakkuk is speaking of history while Nahum is speaking of the future! The heading in the L.D.S. edition of the Bible labels this chapter as a description of "the burning of the earth at the Second Coming." Hence, the similarity of past and futgure cataclysms is reinforced in the writings of Habakkuk and Nahum.

A rather complete picture of a past catastrophe was recreated in this chapter from the obscure book of Habakkuk. Hopefully it has added to our knowledge of the events of the Exodus and of the days of Joshua, and created a more complete picture of catastrophe.

There are still other, better known prophets from the Old Testament who have more to say about past and future catastrophes. In fact, an understanding of past catastrophes allows a marvelous insight into the writings of all the prophets—especially those which are the most enigmatic.

". . . in the days that the prophecies of Isaiah shall be fulfilled men shall know of a surety, at the times when they shall come to pass."

2 Nephi 25:8

3

The Writings of Isaiah

Your country is desolate, your cities are burned with fire . . .

Except the Lord of hosts had left unto us a very small remnant, we should have been as Sodom, and we should have been like unto Gomorrah. (Isaiah 1:7, 9.)

Isaiah described the effects of the "commotion" in the days of Uzziah, King of Judah. To further underscore the extent and the nature of the devastation, he added:

> Therefore is the anger of the Lord kindled against his people, and he hath stretched forth his hand against them, and hath smitten them: and the hills did tremble, and their carcasses were torn in the midst of the streets. (Isaiah 5:25.)

Times of trouble

According to Velikovsky, the days of Isaiah were troubled times for planet Earth. Beginning in the eighth century B.C., Mars menaced the Earth from time to time as the orbits of the two planets converged and parted on at least three occasions (all in the lifetime of Isaiah): once in the days of Uzziah (747 B.C.), again on the day when Ahaz was buried (721 B.C.), and again in the days of Hezekiah (687 B.C.). Velikovsky painted a fascinating picture of the life and times of Isaiah, and his insights are truly remarkable. They are also extremely useful to the student of the scriptures.

Velikovsky's reconstruction of ancient history (based on his dating of past cataclysms) reveals remarkable facts and details relating to Israelite history—from the time of the Exodus, down through the division of Israel, to the days of Isaiah, and on into the time of the Babylonian captivity. These periodic cataclysms in Isaiah's lifetime, attributed by Velikovsky to Mars, afflicted all of mankind for at least sixty years, perhaps more. Therefore, Velikovsky maintained that what Isaiah recorded was his warning to fellow Israelites of the catastrophes to befall them in that day and age.

Past and future

This appears to be a correct conclusion as far as it goes; but there is more to Isaiah's prophecies that Velikovsky did not acknowledge. Not only did Isaiah predict events to occur in his day, he foretold events to occur in the future—far beyond his day. For example, he penned a prophecy of the coming of the Messiah in the meridian of time (Isaiah 53). Isaiah also made reference to a radical change in the topography in and around Jerusalem, that he specifically predicted would occur in the last days. He wrote:

> And it shall come to pass in the *last days*, that the mountain of the Lord's house shall be established in the tops of the mountains, and shall be exalted above the hills . . . (Isaiah 2:2, italics added.)

Since there is no record of Jerusalem being elevated or "exalted" during the catastrophes that befell her in the days of Isaiah, nor at any time since then, it appears that the time when that city will experience the predicted change in geography is still in the future. (See *And The Moon Shall Turn To Blood*, Chapter 8, for a thorough examination of the changes prophecied to occur in and around Jerusalem.)

In conjunction with the above statement, Isaiah made an interesting reference to the Exodus, regarding the future gathering of Israel. He wrote:

> And the Lord shall utterly destroy the tongue of the Egyptian sea; and with his mighty wind shall he shake his hand over the river, and shall smite it in the seven streams, and make men go over dryshod.
>
> And there shall be an highway for the remnant of his people, which shall be left, from Assyria; like as it was to Israel in the day that he came up out of the land of Egypt. (Isaiah 11:16.)

The Lord had obviously shown Isaiah that an event would occur in the last days in conjunction with the gathering of the House of Israel that would be similar to the parting of the waters during the Exodus. The above quote, taken together with Isaiah's statement "that the mountain of the Lord's house shall be established in the top of the mountains," seems to indicate that we may expect to see some rather graphic changes in the topography of the region in and around the Middle East.

It seems that Isaiah's predictions not only concerned events in his day but also extend to events that were to occur far beyond those ancient days of cosmic upheaval in which he lived and prophesied.

Multiple fulfillment

Isaiah's prophecies of catastrophes, as with other principles of truth he taught, can apply to many different times and places throughout history. Nephi and Jacob applied the teachings of Isaiah to their people, the Nephites, using the same statements (2 Nephi 6:5 and 1 Nephi 19:23.) If the circumstances surrounding an event or situation in modern times are similar to another event recorded in the scriptures, then what was written concerning the ancient event or situation can also be applied to its modern counterpart:

> A major difficulty in understanding the book of Isaiah is his extensive use of symbolism, as well as his prophetic foresight and literary style; these take many local themes (which begin in his own day) and extend them to a latter-day fulfillment or application. Consequently, *some prophecies are probably fulfilled more than one time and/or have more than one application.* (The Bible Dictionary, *King James Bible,* L.D.S. Version, p. 707, italics added.)

Hence, descriptions penned by Isaiah of catastrophes to be caused by intersecting planets in modern times, will closely resemble similar events of antiquity. For that reason, Isaiah's descriptions of calamities can apply to events that are yet in our future and may be likened to them.

Because of the prophetic nature of Isaiah's writings (that deal with his day or ours), it should be very enlightening to go beyond Velikovsky's analysis of Isaiah, to examine what Isaiah had to say about cataclysmic events in both time frames.

The pillar of the cloud

In chapter 4, Isaiah described the condition of Zion and Jerusalem during the Millennium (after the destructions and changes). He made this fascinating statement:

> And the Lord will create upon every dwelling place of mount Zion, and upon her assemblies, a cloud and smoke by day, and the shining of a flaming fire by night: for upon all the glory shall be a defence. (Isaiah 4:5.)

Notice how similar that description by Isaiah of a phenomenon seen to exist during the Millennium parallels that made by Moses when he described this striking phenomenon of the Exodus:

> And the Lord went before them by day in a pillar of a cloud, to lead them the way; and by night in a pillar of fire, to give them light; to go by day and night
>
> . . . and the pillar of the cloud went from before their face, and stood behind them: And it came between the camp of the Egyptians and the camp of Israel . . . so that the one came not near the other all the night. (Exodus 13:21, 14:19-20.)

Latter-day prophets have also described seeing this phenomenon in visions of future events. Elder Wilford Woodruff described a vision of the founding of the New Jerusalem:

> I seemed to be standing on the west bank of the Missouri River opposite the City of Independence . . . . I then saw a short distance from the river twelve men dressed in the robes of the temple standing in a square . . . I saw myriads of angels hovering over them and around about them and

> also *an immense pillar of a cloud hover over them. . . . and all the time I saw the great pillar of a cloud hovering over the place.* (*Wilford Woodruff Journals,* June 15, 1878, italics added.)

According to Velikovsky, the pillar of a cloud seen during the Exodus was a result of the appearance of the comet/planet Venus standing vertically on the horizon. The appearance of the same phenomenon in prophecies of the future indicates that a very similar condition will likely prevail. The fact that this pillar is a permanent feature hovering over Zion, standing above the Earth, may be an indication that it has joined Earth in its orbit around the sun. There is much to indicate that the orbits of other planets in our solar system and the orbit of the Earth will be changed substantially in coming near-collisions such that the Earth will be joined in its path through space. The overtaking of and "docking" with Earth by another body or bodies will, in fact, cause some of the prophecied destructions. (See *And The Moon Shall Turn To Blood,* Chapter 9, pp. 89-95.)

Mysterious animals? armies?

The following are a particularly puzzling set of verses from Isaiah:

> And he will lift up an ensign to the nations from far, and will hiss unto them from the end of the earth: and, behold, they shall come with speed swiftly:
>
> None shall be weary nor stumble among them; none shall slumber nor sleep; neither shall the girdle of their loins be loosed, nor the lachet of their shoes be broken:
>
> Whose arrows are sharp, and all their bows bent, their horses' hoofs shall be counted like flint,

and their wheels like a whirlwind:

Their roaring shall be like a lion, they shall roar like young lions: yea, they shall roar, and lay hold of the prey, and shall carry it away safe, and none shall deliver it.

And in that day they shall roar against them like the roaring of the sea; and if one look unto the land, behold darkness and sorrow, and the light is darkened in the heavens thereof. (Isaiah 5:26-30.)

What was Isaiah describing here? On the one hand, he appears to be talking about an army when he says "Whose arrows are sharp, and all their bows bent, their horses' hoofs shall be counted like flint, and their wheels like whirlwind." On the other hand he seems to be talking about "young lions; yea they shall roar, and lay hold of the prey, and shall carry it away safe, and none shall deliver it." Obviously, Isaiah attempted to describe something by using these metaphors, but what was he talking about—an army? animals? or something else?

Perhaps the last verse will help clarify the reference: ". . . and if one look unto the land, behold darkness and sorrow, and the light is darkened in the heavens thereof." Darkened heavens is one of the primary signs of a catastrophe, according to Velikovsky. Could it be that Isaiah was speaking of a natural phenomenon?

In light of the evidence from the last verse, notice that the first verse speaks of "an ensign to the nations. . . ." "Ensign" simply means "sign," and the great sign of the Exodus was the comet/planet that caused it all. Is it possible that Isaiah was speaking of such an event?

Isaiah also seems to be describing the same phenomena in another chapter.

> The noise of a multitude in the mountains, as of a great people; a tumultuous noise of the doms of nations gathered together; the Lord of hosts mustereth the host of the battle. (Isaiah 13:4.)

The initial impression of this vision is that of a gathering of "all nations" to do battle. But the subsequent verses give some added clarification:

> They come from a far country, *from the end of heaven,* even the Lord, and the weapons of his indignation, to destroy the whole land.
>
> Howl ye; for the day of the Lord is at hand; *it shall come as a destruction from the Almighty.* (Isaiah 13:5-6, italics added.)

The Lord of hosts

This, then, is no earthly army. These are God's agents of destruction—a heavenly army of objects from space. The comparison to animals is natural, since their appearance would resemble that of various animals, and the comparison to an army is most appropriate, since they would be a destroying host mustered by the Lord. Lest the reader doubt this interpretation of what Isaiah was describing here, examine a few of the remaining verses from this same chapter:

> Behold, the day of the Lord cometh, cruel both with wrath and fierce anger, to lay the land desolate: and he shall destroy the sinners thereof out of it.
>
> For the stars of heaven and the constellations thereof shall not give their light; the sun shall be darkened in his going forth, and the moon shall not cause her light to shine.
>
> Therefore I will shake the heavens, and the

> earth shall remove out of her place, in the wrath of the Lord of hosts, and in the day of his fierce anger. (Isaiah 13:9-10, 13.)

Can there be any doubt that the army of the Lord described by Isaiah is not an earthly army, but a heavenly one? The metaphors that Isaiah used here were common to the ancients. He was describing natural phenomena in a manner that would have been unmistakable to the people of his time. Joel and the Apostle John both wrote of this same phenomenon. It is the army of God—the hosts of heaven. Isaiah is describing the type of pehnomenon that terrorized mankind quite often during periods of catastrophe. The appearance and actions of these bodies gave rise to some of the most extravagant and hard to understand metaphors and similes in the Bible. (See Chapter 4 for an in-depth analysis of this phenomenon.)

Key to Isaiah

This catastrophic perspective of ancient events is unique, and allows us to better understand much more of the writings of Isaiah. For example, examine the following verses in Isaiah:

> Fear, and the pit, and the snare, are upon thee, O inhabitant of the earth.
>
> . . . for the windows from on high are open, and the foundations of the earth do shake.
>
> The earth is utterly broken down, the earth is clean dissolved, the earth is moved exceedingly.
>
> The earth shall reel to and fro like a drunkard, and shall be removed like a cottage . . . (Isaiah 24:17-20.)

Clearly, Isaiah was describing a catastrophic event of epic proportions. Now notice the next verse with regard to the "hosts of heaven" spoken of earlier:

> And it shall come to pass in that day, that the Lord shall punish the host of the high ones that are on high, and the kings of the earth upon the earth. (Isaiah 24:21.)

On the one hand Isaiah wrote that the Lord would bring His army, His "strong people," His "terrible nations" from the heavens to punish mankind, then on the other he declared that the Lord would drive off the "terrible ones" that plague the Earth:

> Therefore shall the strong people glorify thee, the city of the terrible nations shall fear thee.
>
> For thou hast been a strength to the poor, a strength to the needy in his distress, a refuge from the storm, a shadow from the heat, when the blast of the terrible ones is as a storm against the wall.
>
> Thou shalt bring down the noise of strangers, as the heat in a dry place; even the heat with the shadow of a cloud: the branch of the terrible ones shall be brought low. (Isaiah 25:3-5.)

As additional evidence of Isaiah's use of the imagery of celestial apparitions, consider the story of the fall of Lucifer in chapter 14. I have previously pointed out that Isaiah not only compared the fall of the king of Babylon with the fall of Satan from heaven, but he compared the fall of those two with the fall of the comet/planet Venus (from the greatest sight in heaven to a tiny star on the horizon). (See *And The Moon Shall Turn To Blood,* Chapter 4, pp. 38-40.) Though this practice may seem odd to us, this was not an unusual thing for Isaiah to do. "Every celestial phenomenon, in fact, was held to have its counterpart in human events . . . ." (*The Ancient Gods,* p. 235.)

Survival instructions

Isaiah gave very good advice to those who would attempt to survive the "indignation" of the Lord:

> Come, my people, enter thou into thy chambers, and shut thy doors about thee: hide thyself as it were for a little moment, until the indignation be overpast. (Isaiah 26:20.)

Logic dictates that the safest place to be during the dust-fall and meteor-fall that initiates some interplanetary encounters is indoors. Isaiah was giving sound advice and guidance. This is not just hyperbole and metaphor, but a statement of fact. The following verses substantiate this:

> For, behold, the Lord cometh out of his place to punish the inhabitants of the earth for their iniquity: *the earth also shall disclose her blood,* and shall no more cover her slain. (Isaiah 26:21.)

And if that is not sufficiently plain, the following verses which allude to the serpent or leviathan myth, also speak of the Earth or the sea as "red wine":

> In that day the Lord with his sore and great and strong sword shall punish leviathan the piercing serpent, even leviathan that crooked serpent; and he shall slay the dragon that is in the sea.
>
> In that day sing ye unto her [the earth], a vineyard of red wine. (Isaiah 27:1-2.)

This entire chapter, as many others in Isaiah, deals with the effects of the coming catastrophes upon the peoples and countries in the Middle East (specifically), as well as the entire world in general.

The above quote also reminds us that in the last days there will again appear in heaven (as was the case anciently) a comet/planet which would appear to be two beings doing battle with one another—one a god of light, the other a serpent, dragon, or "leviathan." It may or may not be the same body that gave rise to the original metaphor anciently, but the future battling apparition will strongly resemble the ancient one, because many of the prophets (including Isaiah and John) speak of seeing it. The "dragon that is in the sea," "leviathan," the "crooked serpent," and "the piercing serpent" are all the same or similar phenomena. Recall that the "serpent" changed its appearance almost from moment to moment during past encounters; hence, over the ages it has been referred to by many names. Also, in repeated encounters it undoubtedly appeared to rise or fall over the horizon, as though it were descending into or coming up out of the sea—hence the reference to "the dragon that is in the sea." It was seen, therefore, in many different positions with relation to the heavens and the Earth.

Isaiah made plain

Many verses in Isaiah seem rather strange in their present context. For example:

> Behold, the Lord hath a mighty and strong one, which as a tempest of hail and a destroying storm, as a flood of mighty waters overflowing, shall cast down to the earth with the hand. (Isaiah 28:2.)
>
> Behold, the Lord maketh the earth empty, and maketh it waste, and turneth it upside down, and scattereth abroad the inhabitants thereof. (Isaiah 24:1.)
>
> And the Lord shall cause his glorious voice to be heard, and shall shew the lighting down of his arm, with the indignation of his anger, and with the flame of a devouring fire, with scattering, and tempest, and hailstones. (Isaiah 30:30.)

Separately, these verses present only a partial picture of what Isaiah was describing. But taken together, these (and a multitude of verses like them) graphically portray the events which would accompany a near-collision of planets.

Isaiah's summary

The following verses from Isaiah seem to be a summary of his vision of the last days. It is, indeed, a sobering assessment when examined from the catastrophe perspective:

> Come near, ye nations, to hear; and hearken, ye people: let the earth hear, and all that is therein; the world, and all things that come forth of it.
>
> For the indignation of the Lord is upon all nations, and his fury upon all their armies; he hath utterly destroyed them, he hath delivered them to the slaughter.
>
> Their slain also shall be cast out, and their stink shall come up out of their carcasses, and the mountains shall be melted with their blood.
>
> And all the host of heaven shall be dissolved, and the heavens shall be rolled together as a scroll: and all their host shall fall down, as the leaf falleth off from the vine, and as a falling fig from the fig tree.
>
> For my sword shall be bathed in heaven: behold, it shall come down upon Idumea [the world], and upon the people of my curse, to judgment. . . .
>
> For it is the day of the Lord's vengeance, and the year of recompences for the controversy of Zion.
>
> And the streams thereof shall be turned into pitch, and the dust thereof into brimstone, and the

> land thereof shall become burning pitch.
>
> It shall not be quenched night nor day; the smoke thereof shall go up for ever: from generation to generation it shall lie waste; none shall pass through it for ever and ever . . . .
>
> Seek ye out of the book of the Lord, and read: *no one of these shall fail, none shall want her mate:* for my mouth it hath commanded, and his spirit it hath gathered them. (Isaiah 34:1-5, 8-10, 16.)

Isaiah's declaration that "no one of these shall fail" is also accompanied by the statement that they will all occur together—"none shall want her mate." This is crucial to the picture of celestial catastrophes. None of these signs occurs singly. They invariably occur in groups or following one another in quick succession, because they all have their origin in one single cause. Velikovsky pointed out the correlation—Isaiah seems to substantiate it.

And so it is with nearly the entire book of Isaiah. The imagery of catastrophe is so strong in the writings of Isaiah, it would be impossible to truly understand all he wrote without a proper understanding of the source of the imagery he used. Because the knowledge of ancient catastrophes was lost, modern scholars and students of the scriptures struggle in vain to fully grasp the meaning of Isaiah. However, once we are familiar with the imagery of past cataclysms, it becomes much easier to understand Isaiah's writings.

"I have cut off the nations; their towers are desolate; I made their streets waste, that none passeth by: their cities are destroyed, so that there is no man, that there is none inhabitant."

Zephaniah 3:6

4

The Prophet Joel, Armies and Creatures

The most striking thing about the book of Joel is its similarity to the book of Revelation, written by John on the isle of Patmos. Although the two books were written by men who lived approximately 600-700 years apart, they bear a remarkable resemblance to one another in many respects. In fact, all the writings of the prophets that deal with events of the last days exhibit the same resemblance. Strangely, once this similarity is clearly perceived, the prophetic books become somewhat repetitious to read. Apparently the prophets down through the ages were each seeing the same things in vision; but instead of each prophet describing what he saw in a unique style or manner of writing, they seem to have used a uniform system of metaphors to describe latter-day events.

Of course, Dr. Velikovsky aptly demonstrated that many of the prophets experienced catastrophic events first-hand—they actually survived the cataclysms of their day. Hence, the events they saw depicted in their visions of the last days, almost paralleled what they had personally experienced. Naturally, they would wish to present to the reader vivid images of what they had both experienced firsthand and seen in vision. Joel, likely, experienced these things both in vision and in person.

Much prophecy

The ancient prophets, apparently, were well versed as to future events. We know that Adam "predicted whatsoever would befall his posterity unto the latest generation. These things were written in the book of Enoch, and are to be testified of in due time." (Doctrine & Covenants 107:56.) The logical assumption is that the records of Adam, Enoch, and the other patriarchs were handed down from father to son until the time when the world was swept clean by the waters of the Flood. Noah, likely, preserved those records, carried them safely through the Deluge, and passed them on to his descendents. Certainly Moses, hundreds of years later, had access to some, if not all of them, since it was he who wrote Genesis by condensing the records of his forefathers into one book (in much the same manner that Mormon compiled the history of his people and condensed it into one book).

The presence of the book of Joel in the Old Testament (along with similar prophetic books, such as Isaiah, Ezekiel, Zechariah, and several others) suggests that there were many books anciently—many more than we have today—that prophesied of the destructions to occur in the last days. Exactly when those ancient records were lost to the world is an unanswerable question, but it may be that all or parts of those ancient records were carefully preserved down through antiquity. It is likely that the aforementioned prophets had access to them.

Prophetic language

The language of prophecy had already become a matter of tradition by the time Joel wrote of his vision. Undoubtedly, Joel was very familiar with those sacred records. When the Lord revealed to him the events to transpire in the last days, Joel realized that the events he was seeing were identical to those he had read in the writings of earlier prophets. In addition—if Velikovsky was correct—Joel had experienced those same types of destructive cataclysms. Hence, we may conclude that he not only learned of catastrophes from the ancient records, but was an eyewitness of celestial catastrophe. Naturally, when he began to set down in writing that which he had seen in vision, he used the familiar, traditional metaphors of prophecy; indeed, he may have felt compelled by prophetic tradition to use the same written expressions.

This perspective of prophecy serves to explain why the language of Peter, or John, or even that of Joseph Smith bears a remarkable resemblance to that of Joel or Isaiah. Though not all the prophets had experienced tremendous natural cataclysms, they all saw them portrayed in vision. Thus, they all had an intimate knowledge of such catastrophes.

Key to prophecy

Adherence to this prophetic tradition is an extremely valuable tool to the student of scripture and prophecy. It provides a consistent descriptive system of interpretation. We have only to learn the key to each metaphor, and we unlock its meaning—ancient or modern.

The key to understanding Joel lies in understanding such

references as:

> For a nation is come up upon my land, strong, and without number, whose teeth are the teeth of a lion, and he hath the cheek teeth of a great lion. (Joel 1:6.)
>
> . . . a great people and a strong; there hath not been ever the like . . .
>
> The appearance of them is as the appearance of horses; and as horsemen so shall they run.
>
> They shall run like mighty men; they shall walk every one in his path . . . (Joel 2:2, 4, 7.)

Nearly all who have attempted to interpret prophecy have determined that these verses in Joel describe a terrible army that will fall upon Jerusalem in the last days. Certainly prophecy speaks of an earthly army composed of many nations that will lay siege to Jerusalem in the last days; but if Velikovsky was right, the prophets also spoke of a heavenly army, not just an earthly one.

Velikovsky quoted extensively from the books of Joel and Isaiah. He drew a fascinating comparison of those books with the text of several of the Vedic Hymns. The Vedic Hymns are poetic writings, held sacred by the Hindus. Together with the Rig-Veda (the oldest and most important of the Hindu sacred books) they are to Hinduism what the Old Testament is to Judaism and Christianity; and Velikovsky convincingly pointed out their similarities and the likelihood that their references to armies and strange appearing creatures from heaven stem from the same source. (See *Worlds in Collision,* Part 2, Chapter 5.)

Joel, John, and Isaiah spoke of mysterious armies and strange creatures. Among other things, the Vedic Hymns also tell of armies and creatures called "Maruts." But according to

Velikovsky, they were all describing the same phenomena.

Hoards of comets

Dr. Velikovsky wrote that in the days of Joel and Isaiah,

> Venus had a tail, considerably shortened since the time it was a comet, but still long enough to give the impression of a hanging flame, or smoke, or attached hair. When Mars clashed with Venus, asteroids, meteorites, and gases were torn from this trailing part, and began a semi-independent existence, some following the orbit of Mars, some other paths.
>
> These swarms of meteorites with their gaseous appendages were newborn comets; flying in bands and taking various shapes, they made an uncanny impression. Those which followed Mars closely looked like a troop following their leader. They also ran along different orbits, grew quickly from small to giant size [as they approached], and terrorized the peoples of the earth. And when, soon after the impact of Venus and Mars, Mars began to threaten the earth, the new comets, running very close to the earth, added to the terror, continually recalling the hour of peril. (*Worlds in Collision*, p. 284-285.)

Velikovsky maintained that the Maruts described in the Vedic Hymns and the descriptions of armies and strange creatures by the prophets Joel, Isaiah, and others were all inspired by the apparitions created during the catastrophes in the days of Isaiah. He wrote:

> The comets, just beginning to whirl, looked like revolving torches or writhing snakes; they assumed the form of spinning wheels, and the

> celestial phantasmagoria appeared like swift chariots; changing their forms, the Maruts look like horses racing along the sky, and then again like a host of warriors, leaping, climbing, irresistible. (Ibid., p. 287.)

The following comparison of the writings of Joel and Isaiah to the Vedic Hymns is but a small sample of the comparison prepared by Velikovsky:

Old Testament	**Vedic Hymns**
A day of darkness and of gloominess, a day of clouds and of thick darkness, as the morning spread upon the mountains: a great people and a strong; (Joel 2:2.)	Even by day the Maruts create darkness. The terrible Marut-host of ever-youthful heroes. All beings are afraid of the Maruts: they are men terrible to behold, like kings.
A fire devoureth before them; and behind them a flame burneth . . . . Nothing shall escape them. (Joel 2:3.)	Like a blast of fire. . . . Blazing in their strength, brilliant like fires, and impetuous.
The appearance of them is as the appearance of horses: and as horsemen, so shall they run. (Joel 2:4.)	At their racings, the earth shakes, as if broken, when on the heavenly path they harness for victory. They wash their horses like racers in the courses, they hasten with the points of the reed on their quick steeds.

The earth shall quake before them; the heavens shall tremble; the sun and the moon shall be dark, and the stars shall withdraw their shining. (Joel 2:10.)

You shake the sky. The terrible ones . . . even what is firm and unshakable is being shaken. When they whose march is terrible have caused the rocks to tremble, or when the manly Maruts have shaken the back of heaven. Hide the hideous darkness, make the light which we long for!

And he will lift up an ensign to the nations from far, and will hiss unto them from the end of the earth: and, behold, they shall come with speed swiftly: None shall be weary nor stumble among them; none shall slumber nor sleep; neither shall the girdle of their loins be loosed, nor the latchet of their shoes be broken: Whose arrows are sharp, and all their bows bent, their horses' hoofs shall be counted like flint, and their wheels like a whirlwind, Their roaring shall be like a lion . . . they shall roar like young lions . . . like the roaring of the sea: and if one look unto the land, behold darkness and sorrow; and the light is darkened in the heavens thereof. (Isaiah 5:26-30.)

These strong, manly, strong armed Maruts do not strive among themselves; firm are the horns, the weapons on your chariot, and on your faces are splendours. They who by their own might seem to have risen above heaven and earth . . . they are glorious like brilliant heroes, they shine forth like foe-destroying youths. They who are roaring and hasting like winds, brilliant like the tongues of fire, powerful like mailed soldiers . . . who hold together like the spokes of chariot-wheels, who glance forward like victorious heroes, who are swift, like the best of horses.

Of the Old Testament accounts, Velikovsky wrote:

> The clouds, the fire, the terrifying din, the darkness in the middle of the day; the fantastic figures on the sky of speeding chariots, running horses, marching warriors; the trembling of the earth, the reeling of the firmament, were visualized, felt, and feared on the shores of both the Mediterranean Sea and the Indian Ocean, for they were not localized disturbances, but displays of cosmic forces in cosmic dimensions. Joel did not copy from the Vedas nor the Vedas from Joel. In more than this one instance it is possible to show that peoples, separated even by broad oceans, have described some spectacle in similar terms. These were pageants, projected against the celestial screen . . . . (*Worlds in Collision*, p. 290.)

Armies and creatures in prophecy

Compare the quote from Velikovsky, the verses from the Old Testament, and the Vedic Hymns, to the following verses from Revelation:

> And the fifth angel sounded, and I saw a star fall from heaven unto the earth: and to him was given the key of the bottomless pit.
>
> And he opened the bottomless pit; and there arose a smoke out of the pit, as the smoke of a great furnace; and the sun and the air were darkened by reason of the smoke of the pit.
>
> And there came out of the smoke locusts upon the earth: and unto them was given power, as the scorpions of the earth have power.
>
> And the shapes of the locusts were like unto

> horses prepared unto battle; and on their heads were as it were crowns like gold, and their faces were as the faces of men.
>
> And they had hair as the hair of women, and their teeth were as the teeth of lions.
>
> And they had breastplates, as it were breastplates of iron; and the sound of their wings was as the sound of chariots of many horses running to battle.
>
> And they had tails like unto scorpions, and there were stings in their tails:
>
> And the number of the army of the horsemen were two hundred thousand thousand: and I heard the number of them.
>
> And thus I saw the horses in the vision, and them that sat on them, having breastplates of fire, and of jacinth, and brimstone: and the heads of the horses were as the heads of lions; and out of their mouths issued fire and smoke and brimstone.
>
> By these three was the third part of men killed, by the fire, and by the smoke, and by the brimstone, which issued out of their mouths.
>
> For their power is in their mouth, and in their tails: for their tails were like unto serpents, and had heads, and with them they do hurt. (Revelation 9:2-3, 7-10, 16-19.)

Though John chose to use the term "locust" to describe the heavenly apparitions, can there be any doubt about what he saw and described? The parallel use of symbols such as horses, lions, serpents, chariots, and battling warriors is convincing. He saw what many other prophets have seen before him, and he described what he saw in traditional, descriptive language. In addition, his descriptions parallel those from other cultures which told of past cataclysmic apparitions.

Apparently, the destructions of the last days will bear a remarkable similarity to ancient cataclysms, and, that we might be forewarned, the prophets wrote plainly. Heretofore, the writings of the prophets have seemed mysterious—even incomprehensible. That was not because Joel or any of the prophets attempted to hide the truth from the reader, but because we who live in the last days have not been schooled in the imagery of Earth's cataclysmic history.

Familiar signs

Of course, Joel speaks of other, familiar signs in conjunction with the menacing armies and creatures:

> The earth shall quake before them; the heavens shall tremble: the sun and the moon shall be dark, and the stars shall withdraw their shining . . . .
>
> And I will show wonders in the heavens and in the earth, blood, and fire, and pillars of smoke.
>
> The sun shall be turned into darkness, and the moon into blood, before the great and the terrible day of the Lord come. (Joel 2:10, 30-31.)

All the above signs are the same phenomena that Velikovsky associated with the near collision of two planets. Joel indicated that there would be several "pillars of smoke." During the Exodus there was but one pillar. Obviously, in the future catastrophe described by Joel, there will be many bodies menacing Earth at once, creating impressive (if not horrifying) displays in the heavens above and on the Earth below!

A prophet's admonition

To those who endeavor to learn of the latter-day events, Joel's opening statement in his first chapter is of considerable interest. He wrote concerning the calamities he was about to describe:

> Hear this, ye old men, and give ear, all ye inhabitants of the land. Hath this been in your days, or even in the days of your fathers?
>
> Tell ye your children of it, and let your children tell their children, and their children another generation. (Joel 1:2, 3.)

Even though the entire book is prophetic, this statement by Joel seems to link the catastrophes he predicted to those of the past: "Hath this been in your days, or even in the days of your fathers?" Such a reference seems to indicate that the prophet understood the similarity of past and future calamities. The admonition to teach each succeeding generation of past and future destructions shows the value the prophet placed on a proper understanding of the nature of such things.

Obviously his directive has not been observed, for modern men do not teach these things to their children. In fact, we have been taught (and we in turn teach) that such things have never happened nor ever could happen! A disregard for the validity of this truth has brought us to a state of ignorance concerning celestial catastrophe. It appears that Peter was quite correct when he declared that in the last days men would be "willingly ignorant" of the truths of Earth's past and future. (2 Peter 3.) If the prophet Joel felt a concern for comprehending these things, and subsequently passing them on, perhaps we who stand to experience these things firsthand should show more concern and make a greater effort to understand what the prophets struggled to pass on—to learn what has been lost.

"Alas for the day! for the day of the Lord is at hand, and as a destruction from the Almighty shall it come."
Joel 1:15

5

The Myopia of Modern Man

The author of *God Is Red,* Vine Deloria, Jr., observed:

> We tend to project present understandings of the world backward into the ancient records and test their credibility, not by what they describe or narrate, but by what we consider reasonable given our present knowledge of the universe. ("Myth and The Origin of Religion," *Pensee,* 1974, IX:45-50, p. 45.)

Modern men and women see themselves as unique. Because we have split the atom, traveled to the moon, cured diseases that once plagued mankind, and probed the vast cosmic reaches with our electronic eyes; because we travel swiftly in cars and airplanes, live in comfortable homes, and eat a variety of food unparalleled in all history; because of our abundance, we believe ourselves to be totally different from our forebears (who would have considered the wealth and ease of our day to be incredible—a kind of paradise). Because we believe we are unique, we tend to take an unnatural view of prophecy.

Modern warfare

Logic would suggest that when the prophets were shown, in vision, the world of our day, they must have seen our unique condition. We logically conclude that they must have seen our cities, trains, automobiles, and airplanes. We also assume that when the prophets saw the great battles and destructions of the last days (which they then described in scripture), they must have been seeing sophisticated, modern, mechanized, and computerized warfare—complete with infantrymen wearing special battle gear and armed with advanced weapons; swift-moving, heavily armed and armored tanks; highly mobile fleets of battleships and aircraft carriers; supersonic, missile-firing jet aircraft; great packs of intercontinental ballistic missiles streaking through the atmosphere; and, last but not least, the mushroom clouds of a nuclear holocaust, with all the attendant destruction, radiation-induced sickness, and death that it would inflict on mankind!

Hence, when we turn to prophetic accounts of the last days, we should expect to find there descriptions of the world as we know it. We believe that seeing an airplane, or a tank, or a ballistic missile would have been baffling to men who lived thousands of years ago; hence, we expect those prophets to use crude or perhaps clumsy analogies in their attempt to describe our marvelous machines.

For example, Revelation 9:3-11 is often interpreted as describing modern jet aircraft. Speaking of the "locusts," John said, "And they had hair as the hair of women. . . ." That phenomenon is attributed by many to the contrails of high-altitude jet aircraft or ballistic missiles. In addition, John wrote, ". . . and the sound of their wings was as the sound of chariots of many horses running to battle." The mention of wings in conjunction with a loud roaring or rumbling sound, immediately evokes an image of modern jets and rockets. (See also Joel 2, Isaiah 5:26-30.)

It is common to hear that the battle of Armageddon, prophesied to occur before the Second Coming of Christ, will be a nuclear holocaust. We hear this most frequently from the pulpit. Today, prophecies of war and destruction are interpreted universally as some kind of fantastic, futuristic man-made event.

For example, Hal Lindsey, in his admirable work, *The Late Great Planet Earth,* considered prophecy in this same light. Wishing to learn the course of future events, Lindsey examined prophecy in order to portray a procession of political and social changes to occur in the last days. When he looked at the devastations prophesied in the Bible, he saw them as being caused by man using his terribly destructive machines. Lindsey summarized with:

> He [John in Revelation] predicts that entire islands and mountains will be blown off the map. It seems to indicate an all-out attack of ballistic missiles upon the great metropolitan areas of the world. . . . All of these verses seem to indicate the unleashing of incredible weapons the world over. . . .
>
> The description of torrents of fire and brimstone raining down upon the Red Army, coupled with an unprecedented shaking of the land of Israel could well be describing the use of tactical nuclear weapons . . . (*The Late Great Planet Earth,* pp. 149, 155.)

In all fairness to Lindsey, it should be pointed out that he did indicate that "this could be a direct judgment from God, or God could allow the various countries to launch a nuclear exchange of ballistic missiles upon each other" (p. 150). Even though he acknowledged that the devastations could be an act of God, Lindsey apparently opted to conclude that these destructions would be instigated by man—caused by nuclear detonations or some kind of "super weapon."

The wrath of God

But the concepts explored in this book make the nature of the prophesied destructions clear: they will be sent of God, not engineered by man. It appears that what the prophets saw and described were natural events, not futuristic warfare. As explained in previous chapters, even the statements from Revelation concerning the strange and mysterious "locusts" has a much more complete and logical explanation than the current, popular explanation.

For example, God said this about the destructions of the last days:

> And I [God] will plead against him with pestilence and with blood . . . an over-flowing rain, and great hailstones, fire, and brimstone. Thus I [God] will magnify myself, and sanctify myself . . ." (Ezekiel 38:22.)
>
> And I [the Lord] will camp against thee round about, and will lay siege against thee. . . .
>
> Thou shalt be visited of the Lord of hosts with thunder, and with earthquake, and great noise, with storm and tempest, and the flame of devouring fire." (Isaiah 29:3, 6.)

Duane S. Crowther in his scholarly work, *Prophecy—Key To The Future,* reaches the same conclusion:

> Many have wondered what the source of the fire will be at the time of Christ's coming in glory. The angel Moroni explained this matter to the Prophet Joseph, saying that it would be the heavenly hosts (and therefore not atomic or hydrogen bombs) who will burn the wicked. (*Prophecy—Key To The Future*, p. 240.)

Through Crowther apparently did not recognize the elements of celestial catastrophe in the scriptures, he correctly assessed the nature of future destructions. Crowther also cited the following statement from the Doctrine and Covenants:

> For the hour is nigh and the day soon at hand when the earth is ripe; and all the proud and they that do wickedly shall be as stubble; and *I will burn them up*, saith the Lord of Hosts, that wickedness shall not be upon the earth . . . (Doctrine & Covenants 29:9, italics added.)

Thus, the Lord has declared emphatically that the destructions to be seen in the last days will be His work, not the handicraft of mankind.

Misinterpretation

Nearly all those who have sought to interpret prophecy have done so through the bias of modern experience. Most evangelists—the so called "millennialists"—have seen modern nuclear warfare as the means of fulfilling the "fire and brimstone" prophecies. Even some Latter-day Saints have misunderstood the prophetic references in the scriptures to natural catastrophes. They cannot be faulted for their attempts to understand the Bible. Such misinterpretation is understandable when we consider how science has discredited the reality of past cataclysms described in the Bible and other ancient documents, thus causing many earnest and sincere scholars to err.

Because we have never experienced the events described by the prophets, we fail to grasp their intended meaning. The prophets' frame of reference was entirely different from our own. When we try to understand prophecy, we usually analyze it in terms of our modern experience, just as Deloria indicated (in the quote that opened this chapter). Instead of interpreting prophecy in terms of our own experience, we must learn to interpret it in terms of the prophets' experiences.

Prophecies of war

In truth, there are prophecies which predict great wars in the last days. There is no doubt about that point:

> . . . and I will bring thee forth, and all thine army, horses and horsemen, all of them clothed with all sorts of armour, even a great company with bucklers and shields, all of them handling swords. Thou shalt ascend and come like a storm, thou shalt be like a cloud to cover the land, thou, and all thy bands, and many people with thee. (Ezekiel 38:4, 9.)

If future destructions follow the form of past destructions, we may expect to see a great deal of warfare among the peoples of the Earth. Velikovsky pointed out that whole nations had migrated from their homelands during past destructions in order to flee the devastation they saw around them. Apparently, the survivors of those ancient catastrophes became extremely war-like and aggressive as they wandered in search of relief from deprivation and hunger. It seems quite likely that identical conditions will exist during and after future cataclysms.

No modern weapons

In reality, it is doubtful that modern, mechanized

warfare would exist after the initial catastrophe. The ability to wage modern warfare, with its complex machinery and high technology, would be obliterated by the first cataclysm. In a matter of moments, the vaunted technology of our time would be utterly destroyed. Manufacture would cease instantly. Delivery systems for food and energy would terminate abruptly. In cataclysmic convulsions of all nature, mankind would be thrown into a struggle for mere survival. With our complex, sophisticated systems destroyed, we would have to learn to exist on a very primitive level. And in order to survive, the wicked would turn to fighting one another for the few remaining resources and commodities. Hence, all the prophecies of the last days describe a law-less, war-like mentality in fallen mankind.

If there is to be wholesale or even limited nuclear war in the future, it will have to occur between the present and the first prophecied catastrophe, or "woe" as John called it. Because the technological works of man will be destroyed in the first cataclysm, it appears that the battles prophesied to occur in the very last days, will be fought as battles were fought anciently—in hand-to-hand combat, with crude weapons. (See Ezekiel 38:4 and Ezekiel 39:9.) Notice the weapons of the army that assaults Jerusalem: shields, bucklers, bows and arrows, hand staves and spears—not rifles, grenades, artillery shells, or tanks.

Interpreting prophecy

As we attempt to interpret prophecy, perhaps it would be well to heed Peter's admonition. We not only have the word of Christ, Peter said, but

> . . . we have also a more sure word of prophecy; whereunto ye do well that ye take heed, as unto a light that shineth in a dark place, until the day star arise in your hearts: Knowing this first,

> that *no prophecy of the scripture is of any private interpretation.* For prophecy came not in old time by the will of man: but holy men of God spoke as they were moved by the Holy Ghost. (II Peter 1:19-21, italics added.)

If modern men and women are to understand the prophetic warnings from God, we must accept the scriptures as describing authentic events. We must not reject them as myth and legend, or interpret them in terms of modern experience. If Velikovsky's assumptions concerning past cataclysms are correct, there can be no mistake about the meaning of the prophets' message.

"Teach ye diligently . . . of things both in heaven and in the earth, and under the earth; things which have been, things which are, things which must shortly come to pass . . . ."

Doctrine & Covenants 88:78, 79.

6

Slowly? or Suddenly?

Undoubtedly, the explanations of the past and the future, presented in this book and its predecessor *And The Moon Shall Turn To Blood,* must seem to be an abrupt departure from current scientific and scholastic thinking—and so they are. But a brief overview of the evolution of the scientific theories supporting or contradicting these explanations may provide additional insight into their validity.

Uniformitarianism

Uniformity is an all-inclusive term used to describe the current nature of scientific thinking. It actually began as a term for use in the field of geology. The theory of uniformity holds that all geologic processes operate in a uniform, consistent manner. That is, all the forces of nature that have made the mountains, valleys, rivers, lakes, continents, and oceans have operated uniformly throughout time. According to the theory of uniformity, we can see those forces that shaped our world operating all around us today. The elements of nature that we experience every day are the creators of our world. Most evident are the wind and the water that are said to eventually erode away even the highest mountains until they are flat plains or sandy beaches. Other forces, seldom perceived or experienced but extremely powerful, slowly push up mountains where once there was a level plain, or those same forces can pull down a landmass until it is buried beneath the ocean.

Of course, as science tells us, all this happens so slowly that man is almost totally unaware of the changes. At the exceedingly slow rate that these changes are taking place today, it would have taken vast amounts of time to make the world we see around us today. So, science reasons that the Earth has been in existence for millions of years.

A scientist who embraces the theory of uniformity is called a *uniformitarian,* and a belief in the theory of uniformity is called *uniformitarianism.* Uniformitarianism is not a religious belief (though it may supplant it); but rather, it is a secular, scientific theory which has been embraced and defended by most of the scientific establishment. Some scientists, like the astronomer Carl Sagan, have gone so far as to declare this theory to be proven fact, when in reality, it is just a theory—no more.

The rise of uniformity

The doctrine of uniformitarianism is a relative newcomer, an outgrowth of the sciences which were born in the Renaissance from the ignorance and superstition of the Dark Ages. Logic and reason, the two cornerstones of modern science, were the tools that intelligent men in the early days of science used in their attempts to overcome the ignorance that prevailed among their contemporaries. But, sadly, their plain logic and careful, rational thinking were often unconvincing to the prevailing political and social powers of their day. Many of those early scientists were treated as fools or charlatans; others were persecuted, and some gave their lives for daring to contradict accepted dogma. Those early scholars and scientists developed a strong resentment for the institutions which persecuted them and resisted their new-found knowledge. Therefore, men of logic and reason began to divorce themselves from any institution that represented the old order of things. One such institution—their major opponent—was religion, whether Catholic or Protestant.

Science rejected religion, the Bible

Hence, the proponents of the infant sciences for the most part rejected religion: its doctrines and its writings. In doing so, they quickly divested themselves of any interest in the recorded "myths" of religion—the scriptures. Many were openly antagonistic toward the Bible. By the time of Darwin in the 1800s, science had relegated the scriptures to the dusty shelves of ancient literature, clearly marked and labeled "myths and legends."

Scientists rejected the Biblical view of creation, maintaining instead that the Earth had been in existence for a much greater length of time than the Biblical account allowed. They reached that assumption on the basis of the following logic: if the forces that alter the Earth today are the same as those that altered it anciently, then the Earth must be very, very old, because those forces change it extremely slowly—almost imperceptibly.

They reasoned that the stratographic record found in the Earth itself was mute evidence of its great age: slice away the side of any hill or mountain to reveal the strata that make it up, and the many layers can easily be seen. There is primarily only one way (they said) to deposit those different layers of material—with water. Scientists reasoned that the ancient layers of sediment must have been deposited in the identical manner that it is being done today. Rivers and streams erode tons of dirt and silt and wash it into lakes and oceans where it is deposited, layer on layer. At the current rate that rivers carry material to the oceans, it would necessarily have taken billions of years to create the layers that can be seen in the Earth's stratographic record.

Some early scientists, like Cuvier, Agassiz, Sedgwick, and Murchison, reasoned that such layering could have happened much more quickly, in cataclysmic inundations—like that of the Deluge. But other scientists would have none of that. Such thought, they reasoned, smacked of the stifling and stultifying dogmas of religion, which they abhorred. Hence, the theories of uniformitarian scientists such as Lyell, Hutton, and Darwin prevailed in a scientific community seeking to divorce itself completely from any religious ideas. Because of the historical animosity that existed between science and religion, scholars and scientists denied that the Biblical record could possibly have anything to do with the world as it really was or had been.

As uniformitarian theories gained more credence, and more discoveries were interpreted according to those theories, science began pushing back the age of the Earth and man's existence thereon further and further. It seemed that each new breakthrough only served to strengthen the uniformitarian position. Increasingly, they postulated the necessity of ever more vast amounts of time in order to allow the gradual processes of nature to bring the world to its present condition.

This state of affairs has continued right up until our time. Uniformity has become more than a theory to the present generation—it is taught as fact. Yet despite its near universal acceptance as fact, it is still only a theory. The monolith of uniformitarianism has stood unrivaled in the modern shrine of scientific thinking for over 150 years, but now some cracks are beginning to appear around its foundations.

Religion joined science

Another interesting phenomenon began to occur about the time that uniformitarianism gained prominence. Men of religion, who had previously espoused the catastrophic concepts found in the scriptures, reversed their point of view. Because of the widespread popularity of the uniformity concepts, they began to see the scriptures as metaphorical accounts rather than statements of fact. In essence, they adopted the rhetoric of science. Apparently this occurred because men of religion felt that the men of science had the correct information. It seemingly became difficult for them to contradict the new ground swell of uniformitarian thought. So, to bring religion into conformity with science, the ancient catastrophic events described in the Bible were interpreted as "myths" or "metaphors" by men of religion. The defenders of the faith had abdicated their positions. Now there appeared to be no one to defend the reality of the scriptures as accurate history. The final result of this doctrinal erosion is evident in our day: we are indoctrinated with uniformitarian principles both by science and by false religion, so that our perception of the Bible is that of a metaphorical, mythical record rather than an accurate account of ancient history, as it was once regarded.

We, therefore, as modern men and women, have received a legacy which teaches us to disbelieve the accuracy and validity of Biblical accounts because they seem so fantastic. Science disbelieves the accounts dealing with phenomena that cannot be tested experimentally—miracles—

and denies the possibility of manifestations that are not in evidence around us today. Hence, when we attempt to read prophecy, which is firmly rooted in the metaphorical imagery of past destructions, we fail to grasp its meaning because we do not understand the origins of the similes and metaphors the ancients employed. We are schooled in uniformitarian thought rather than catastrophic thought, and so we are unprepared to understand even the simplest reference to catastrophe, whether in the scriptures, mythology, or the records of any ancient culture. And because the language used by the prophets to describe ancient destructions is the identical language they used to describe future events, we are left ignorant of the correct message of prophecy. The only way to remedy that situation is to learn the truth about ancient catastrophes. To correctly understand the past is to know the future.

Catastrophism

Our present position in history allows us an excellent vantage point from which to examine the growth of uniformitarianism. But it also lets us see that the theory of uniformity grew at the expense of another, equally valid theory of change—catastrophism.

The theory of *catastrophe* holds that major change is sudden and cataclysmic, rather than slow and gradual—that the topography of the Earth has been altered suddenly, from time to time, during incredible paroxisms of all nature. During these cataclysmic episodes, the ordinary forces which at present are slowly changing our world suddenly accelerate a million-fold, bringing about dramatic change in a short period of time. Catastrophists maintain that during such events mountains erupt from level plains, land-masses are swallowed whole by the oceans, rivers abruptly change courses, cold polar regions exchange places with tropical regions, and all nature goes on a rampage which devastates mankind and his environment—all in the short space of a few hours, days or weeks.

The theory of catstrophism puts an entirely new light on the topic of geology and the age of the Earth. If the rate of change which we experience today was not always constant—if it accelerated dramatically from time to time during past catastrophic episodes—then Earth could have reached its present condition much more rapidly than predicted by the theory of uniformity. That conclusion more nearly coincides with the scriptural record of Earth's history. Hence, the catastrophists, who wish to be thought of as scientists, not religionists, sometimes find themselves in the uncomfortable position of being on the side of the religious fundamentalists who hold the scriptures to be accurate history.

The Mammoth and the Ice Age

A good illustration of the conflict between the theories of catastrophe and uniformity may be seen in the debate over the problem of the extinction of the great mammoths during the last ice age. The uniformitarian view of the last great ice age holds that for a prolonged period of tens of thousands of years, the Earth underwent a cooling process wherein ice and snow slowly accumulated at the poles. The average temperature slowly, virtually imperceptibly, dropped (so the theory goes) and more and more ice and snow accumulated in these great arctic and antarctic glaciers. Eventually, as the ice age progressed, the vast glaciers of accumulated ice and snow pushed their way across continents toward the more temperate zones on the Earth. As they did so, the mammoths were unable to adapt to the colder climate, and caught in the advancing ice and snow, eventually they died out.

Then, over time, the Earth began to warm again, and the snow and ice accumulated less rapidly. Eventually, the great glaciers covering the continents began to melt and retreat to the poles, leaving the continents scarred and gouged with great depressions in some places. Melting ice and snow filled these places with water to form lakes and river valleys.

But there are some problems with that theory that will become evident as we examine the point of view of the catastrophists.

A better alternative

The catastrophists believe it all happened quite differently. Although they are certainly not in total agreement with one another as to the exact scenario, nearly all catastrophists point to extraterrestrial bodies as the cause of Earth's ice ages, including the last one.

Velikovsky felt that the ice ages, as defined, were a totally fictitious creation. He held that in past planetary close encounters, the position of Earth's axis was changed. As a result, the poles changed position, too. Hence, a region of the globe that had been warm, verdant and semi-tropical suddenly fell, in a matter of moments, within the newly-formed, frigid polar regions as the Earth "tipped." Plants and animals alike, including the mammoths, quickly would have frozen to death. The so-called ice ages came and went as the location of the poles changed during past catastrophes.

Velikovsky also pointed out a glaring flaw in the uniformitarian theory of the ice ages. In order for ice and snow to be deposited in prodigious quantities on the polar ice caps, it first would be necessary for the Earth to undergo extensive heating. This would facilitate evaporation and suspension in the atmosphere of *massive* amounts of water vapor sufficient to create the great ice-sheets. The heating would have to be followed quickly by rapid cooling, so that the enormous amounts of water necessary to create the great ice sheets of the ice ages could precipitate onto the poles as snow. Because heating is necessary to suspend tremendous amounts of water vapor in the atmosphere before it can precipitate on the poles, Velikovsky maintained that the uniformitarian ice age model is not functional.

Velikovsky, along with other catastrophists like Donald W. Patten, believed that ice was ejected from another body in the solar system which was then proximate to the Earth. Patten convincingly pointed to the icy rings of Saturn as evidence of past ice cataclysms in the solar sysetem. He claimed the ice crystals deposited around Earth were electrically charged and thus followed the planet's magnetic flux lines to the poles, where they precipitated suddenly from space as tons of super-cooled ice. The ice from space—thus precipitating only at the poles—brought temperatures approaching absolute zero (-459 degrees Fahrenheit) to that part of Earth's surface. Any plant or animal unlucky enough to be caught beneath the rapidly accumulating ice-fall was immediately quick-frozen. Because the native habitat of the mammoths suddenly had been thrust into the polar region by the newly-tipped Earth, all of those prehistoric creatures (along with all the other plants and animals common in their habitat) were extinguished within moments, buried in a deluge of ice falling from space. (See *The Biblical Flood and The Ice Epoch*, Chapter 6.)

According to specialists in the new technology of quick frozen or freeze-dried food, the great mammoths found in the arctic ice, their flesh intact, could only be so well preserved by such a method. Normal freezing temperatures common in today's polar regions would have been too slow a method to preserve the flesh of the mammoths—large ice crystals would have had time to form within the flesh, damaging it, and some of the mammoths' flesh would have spoiled before it would have frozen. Quick freezing, however, is almost instantaneous and nearly uniform. In addition, quick freezing occurs so rapidly that large ice crystals cannot form, thus better preserving the animals. This can only be done at temperatures near absolute zero.

Therefore, what the theory of uniformity estimates to be a prolonged event—of tens or hundreds of thousands of years in duration—becomes a cataclysmic event that is quite sudden and endures only a tiny fraction of the time previously thought necessary. Because of such foreshortening, the vast amounts of time supposedly required for geologic processes to occur, simply did not exist. This explanation serves to illustrate how theories of catastrophe dramatically reduce the time considered necessary for events in evidence in Earth's geologic history to take place.

Earth's age

The age of the Earth has long been a topic of contention between science and Christianity. The scriptures declare that the Earth was created in "6 days." Uniformitarian science has often ridiculed Christians for being so naive as to believe such simple-minded dogma. Pragmatic, scientific observation and deduction, based on the uniformity theory, has led men to believe that the Earth is very old, therefore the biblical claim that the Earth was created in only six days must be fanciful mythology. However, in a catastrophic cosmic event, all geologic processes would be sped up a thousand fold. Sediment deposition, land-mass settling, and mountain formation would occur in a much shorter period of time. The theory of catastrophism thus radically alters the picture of Earth's creation and geologic history.

If catastrophe theories are correct and if, as Peter says, one day to God is as a thousand years to man, the Earth could have easily come into existence—complete with its geologic record, ready for human habitation—in an amount of time comparable to the biblical period of creation—6,000 years! Therefore, the theory of catastrophe gives great strength and validity to the scriptural claim of Earth's age by providing the means whereby Earth could have been created in a relatively short period of time, while still conforming to the evidence in the geologic record.

Catastrophism rejected

Historically, catastrophism was eventually rejected for two reasons. First, it was the theory of change espoused by the scriptures. Second, none of the processes necessary for catastrophic change appear to be in evidence around us today. Science is very pragmatic. If it cannot test a theory by observing it in operation, it doubts the validity of that theory.

New evidence

But as with all truth, the truth of catastrophism cannot be held in check for long. In the past twenty years more and more evidence supporting catastrophism—and thereby discrediting uniformitarianism—has been discovered, particularly in conjunction with man's exploration of space. As the marvelous machines we have sent into space report to us on the conditions that prevail there, more and more is being said and written by the scientific community about evidence of *great planetary catastrophes in our own solar system!* The recent discoveries that have occurred as we moved into the space age have so radically and dramatically changed our concepts of the solar system that any astronomy textbook written more than ten years ago now reads like a text from the Dark Ages.

These discoveries are coming so fast and so frequently that scientists themselves are hard pressed to keep abreast of all the changes. Space probes have revealed Jovian moons (some nearly the size of our Earth) with cracks in their crusts several times larger than the Grand Canyon. One such crack or rill on Mars is as long as the United States is wide! Such fractures testify that tremendous external, catastrophic forces were once applied to those orbs. Enormous craters, like the huge impact crater on Saturn's moon Mimas, are mute evidence of meteors large enough to shatter planets. The rings of Saturn and the great asteroid belt between Mars and Jupiter are striking evidences of cataclysms that may have dismembered a moon or perhaps an entire planet in the past.

Increasingly, evidence gathered from space has forced scientists to re-think their cherished theories of uniformity. In 1973, Harold C. Urey wrote: ". . . it does seem possible and even probable that a comet collision with the Earth destroyed the dinosaurs and initiated the Tertiary division of geologic time." ("Cometary Collisions and Geological Periods," *Nature*, 242, March, 1973, p. 32.) Later, in 1976, Adrian J. Desmond observed:

> According to the latter-day catastrophists, the dinosaurs exited with the most spectacular bang since Creation, and, the geologist's lingering aversion to cataclysms notwithstanding, it is becoming difficult to disagree. . . . Supernova or otherwise, whatever racked the planet 70 million years ago brought to an abrupt end the Mesozoic world order and closed a major chapter of earth history. . . . The magnitude of the devastation cannot be underestimated . . . Whatever the nature of the event that we are dealing with, it was of cataclysmic proportions. (*The Hot Blooded Dinosaurs*, pp. 195-96.)

Most recently, in 1981, a team of scientists from the University of California (Berkeley), headed by geologist Walter Alverez, theorized that the dinosaurs became extinct when the Earth was impacted by a giant comet or asteroid—enveloping Earth in a thick cloud of dust, blocking out the sun, and thereby destroying all plant life. Because of suddenly dwindling foodstuffs, animals weighing over 75 pounds perished. Their theory is based on hard, empirical data from analysis of samples from the Earth's geological record.

Changing thought

Such a premise would have been laughable as recently as twenty years ago—*but not now.* The weight of evidence is falling increasingly on the side of catastrophism. The previous example of changing thought is by no means unique, but it would necessitate a book much larger than this to explore the changes in scientific thinking that have taken place recently.

Though the science establishment vehemently denies that the theory of catastrophe has any place in current scientific thought, individual scientists seem to be turning, increasingly, to the tenets of catastrophism in order to explain the newest discoveries. It would seem that the theory of uniformity is dying a slow, suffocating death, while the fresh breezes of discovery are breathing new life into the revived theory of catastrophe.

In this chapter we have seen how and why the concepts of catastrophe were totally displaced by the concepts of uniformity. It appears obvious why ideas of celestial catastrophes seem strange and unfamiliar to us. It is evident that we have been insulated from the truth about past catastrophes by the pervasive theory of uniformity. Evidently, we are ignorant of the message our predecessors struggled to communicate—that which was common knowledge among our distant ancestors; and because of that ignorance we misinterpret and misunderstand the few ancient records that have been preserved—the scriptures, for example. But what do the ancient records, other than the scriptures, have to say about past catastrophes? Is there evidence in those records to support Velikovsky's theories?

". . . were it not for these things, which have been kept and preserved by the hand of God, that we might read and understand of his mysteries . . . we should have been like unto our brethren, the Lamanites, who know nothing concerning these things, or even do not believe them when they are taught them, because of the traditions of their fathers, which are not correct."

Mosiah 1:5

7

Evidence from Ancient Records

Velikovsky wrote this about the abundance of ancient evidence attesting to the truth of past inter-planetary catastrophies:

> In hymns, in prayers, in historical texts, in philosophical discourses, in records of astronomical observations, but also in legend and religious myth, the ancients desperately tried to convey to their descendants, ourselves included, the record of events that took place in circumstances that left a strong imprint on the witnesses. There were physi-

> cal upheavals on a global scale in historical times; the grandiosity of the events inspired awe. From the Far East to the Far West—the Japanese, Chinese and Hindu civilizations; the Iranian, Sumerian, Assyrian, Babylonian, Hitto-Chaldean, Israelite and Egyptian records; the Etruscan, Attic and Roman theogonies and philosophies; Scandinavian and Icelandic epics; Mayan, Toltec and Olmec art and legends—all, with no exception, were dominated by the knowledge of events and circumstances that only the most brazen attitude of science could so completely disregard. (*Velikovsky and Establishment Science,* pp. 5-6.)

Man's forgotten past

And why, we might well ask, has modern man discarded the myths and legends that told of great calamities in the past? Ignatius Donnely, an early catastrophist, made this insightful comment:

> Civilization brings with it a contempt for everything which it can not understand; skepticism becomes the synonym for intelligence; men no longer repeat; they doubt; they dissect; they sneer; they reject; they invent. If the myth survives this treatment, the poets take it up and make it their stock in trade: they decorate it in a masquerade of frippery and finery, feathers and furbelows, like a clown dressed for a fancy ball; and the poor barbarian legend survives at last, if it survives at all, like the Conflagration in Ovid or King Arthur in Tennyson—a hippopotamus smothered in flowers, jewels, and laces.
>
> Hence we find the legends of the primitive American Indians adhering quite closely to the

> events of the past, while the myths that survive at all among the civilized nations of Europe are found in garbled forms, and only among the peasantry of remote districts. (*Ragnarok*, p. 117.)

Mankind's amnesia

Thomas A. Parry (Ph.D., University of California, Berkeley, and the Graduate Theological Union, Berkeley) felt (as did Velikovsky) that the trauma suffered by the survivors of ancient catastrophes was so intense that a form of mass amnesia afflicted all of mankind; so that after having suffered the horror of the last great cataclysms of the eighth and seventh centuries B.C.

> man apparently began to look for more tangible sources of security, and even began to deny that the celestial world was, or could ever have been, unstable or inconstant. It is generally considered that, in Western thought, such a watershed took place in the work of Plato and Aristotle. The pre-Socratic philosophers, Heraclitus most notably, had founded their thought on a frank recognition of the transiency and inconstancy of the world. Plato, in some senses, bridges the past recognition that there has been dramatic inconstancy in the past, in the quest for a basis for a source of unquestioned stability. At virtually the same time as he acknowledges that there have been great cataclysms in the past, Plato urges that those who do entertain such dangerous beliefs be severely punished. (*Kronos*, Vol. I, No. 1, pp. 14-15.)

Hence, the correct picture of past cataclysms was suppressed—*but not totally forgotten.* Prior to the emergence of modern science and uniformitarianism, many early scholars espoused the cataclysmic theory of change. For the most part, they learned of the validity of catastrophism from the geologic record in the Earth and from written accounts of cataclysms and great changes in our solar system anciently. Those accounts were already ancient then, and many are no longer in existence today. In explicit terms they told of great changes in our solar system anciently.

Myths? history!

A cursory perusal of the writings of those early scholars, who had access to the ancient records and commented on them in their own writings, reveals much about the history of the Earth and the solar system. Thus, their writings serve as a bridge to connect us with the distant past and inform us of what the ancients had written about their time—mist-shrouded antiquity.

I will cite only a few examples for the sake of the reader rather than cite the exhaustive research into ancient writings that others have done. With a modest amount of effort and the new-found perspective of catastrophism initiated by Velikovsky, the reader can discover for himself the overwhelming evidence for catastrophism found in ancient texts.

The first example comes from Plato's *Timeaus,* wherein he makes the remarkable statement:

> There have been, and will be hereafter, many and divers destructions of mankind, the greatest by fire and water, though other lesser ones are due to countless other causes. Thus the story current also in your part of the world, that Phaethon, child of the Sun, once harnessed his father's chariot but could not guide it on his father's course and so burnt up everything on the face of the earth and was himself consumed by the thunderbolt—this legend has the air of a fable; but *the truth behind it is a deviation of the*

> *bodies that revolve in heaven round the earth and the destruction, occurring at long intervals, of things on earth by a great conflagration.* (*Timeaus*, 22 C, D, p. 15.)

Examining Ovid's account of the story of Phaethon which Plato was referring to, we find an incredible account of a past cataclysm:

> . . . the earth in flames,
> Mountains touched first, hills, plateaus, plains,
> The dry earth canyon-split, the fields spread white
> In ashes; trees, leaves were branches of the flames
> While miles of grain were fuel for their own fires—
> But these were the lesser losses I regret.
> The great walled cities perished; nations fell.
>
> . . . Then Libya
> Became a desert where wild flames ate the dew,
> Even the rain that swept across her grasses;
>
> . . . And fire tossed on Babylon's Euphrates,
> Fire on Orontes and rapid Thermondon
> And on the Ganges, Phasis, and the Hister;
> Alpheus boiled and banks of Spersheos
> Were streamed with fire while the golden sands
> Of Tagus melted in flames.
>
> . . . Nile ran in terror to the end of earth
> To hide its head which now is still unseen;
> Its seven mouths fell open, filled with dust,
> The seven beds scorched dry, the same fate falling
> On Thracian rivers, Hebrus and Strymon,
> And rivers of the West, Rhine, Rhone, and Po—
> Tiber, whose promise was to rule the world.
> Earth-wide, great canyons opened to the sun,
> And to the fears of Pluto and his queen,
> The sky shed glares of light throughout their kingdom;
> The seas shrank into sand and from their waters

> The hidden mountains rose and Eastern islands
> Came where the waves had vanished.
> (*The Metamorphoses* II, pp. 62-64.)

Remember, Plato wrote that the above account is *an actual description,* not a myth, of what happens to the Earth and the Heavens when celestial bodies go out of their courses.

In Shakespeare's *Troilus and Cressida* (1602), for example, we find Ulysses saying:

> . . . but when the planets
> In evil mixture to disorder wander,
> What plagues, and what portents, what mutiny,
> What raging of the sea, shaking of earth,
> Commotion in the winds, frights, changes, horrors,
> Divert and crack, rend and deracinate
> The unity and married calm of states
> Quite from their fixture!

It would appear that even as late as the time of Shakespeare, records still existed that explicitly tied earthly calamities to wandering planets, and some men still believed them to be accurate records rather than myths.

Hippolytus quoted Democritus, one of the great scholars of antiquity, as having written:

> The worlds are unequally distributed in space; here there are more, there fewer; some are waxing, some are in their prime, some waning: coming into being in one part of the universe, ceasing in another part. The cause of their perishing is collision with one another. (*The Refutation of All Heresies,* I, Chapter XI.)

Of course, modern scholars claim that men like Democritus were lacking in the proper knowledge of the organization of the universe, but it may well be that they learned of the realities of the past from documents that contained eyewitness accounts of cataclysmic changes within our solar system.

St. Augustine, in his *The City of God,* wrote this most remarkable account of a cosmic event:

> I shall, therefore, cite a passage taken, word for word, from a work of Marcus Varro, his *Origins of the Roman People:* "There was once a remarkable portent in the sky; for Castor reports that the planet Venus—Vesperugo, as Plautus calls it, and Homer's 'Hesperus, loveliest of stars'—by a marvel such as never occurred before or since, once changed its color, size, shape, and course. According to the famous astronomers, Adrastos of Cyzicus and Dion of Naples, this occurred during the reign of King Ogyges." Now, so careful a writer as Varro would not use the word "portent" in this connection unless the phenomenon seemed contrary to nature. A portent means, in ordinary parlance, "something contrary to nature," although, in fact such happenings are not really contrary to nature, for the simple reason that nothing that happens by the will of God can be "contrary to nature." The "nature" of any particular created thing is precisely what the supreme Creator of the thing willed it to be. Hence, a portent is merely contrary to nature as known, not to nature as it is. Now, an immense number of portents are recorded in pagan histories. Let us, however, keep to the one point at present in debate. Nothing, I take it, has been so determined by the fixed and unchanging laws of the Author of the nature of the heavens and the earth as the ordered courses of the stars. Nevertheless, the moment He who has supreme authority and power over every-

> thing He created willed a change, then the greatest and brightest of all the planets changed its color, size, shape, and, what is still more remarkable, the law and order of its course. Then, if ever, God played havoc with the rules—if there were any—of the astronomers, those written canons by which they calculate—unerringly, as they imagine—the past and future movements of the stars, the very rules in virtue of which they were confident in declaring that nothing like it ever happened to the Morning Star [Venus] before or since. (Aurelius Augustinus, *The City of God,* [De Civitates Dei], XXI, 8.)

A statement by Pliny, the Roman naturalist and writer, is extremely significant from the catastrophist point of view:

> The theory of the Babylonians deems that even earthquakes and fissures in the ground are caused by the force of the stars that is the cause of all other phenomena, but only by that of those three stars (planets) to which they assign thunderbolts. (Pliny II, 86, as quoted in *Worlds in Collision,* p. 282.)

Hesiod, the Greek poet, wrote of Typhoeus (Typhaon, or Typhon), a serpent-like monster:

> But beneath his immortal feet vast Olympus trembled, as the king uprose and earth groaned beneath. And the heat from both caught the dark-colored sea, both of the thunder and the lightning, and fire from the monster, the heat arising from the thunder-storms, winds, and burning lightning. And all earth, and heaven, and sea, were boiling; and huge billows roared around the shores about and around, beneath the violence of the gods; and unallayed quaking arose. Pluto trembled, monarch

> over the dead beneath; and the Titans under Tartarus, standing about Cronus, trembled also, on account of the unceasing tumult and dreadful contention. But Jove, when in truth he had raised high his wrath, and had taken his arms, his thunder and lightning, and smoking bolt, leaped up and smote him from Olympus, and scorched all around the wondrus head of the terrible monster.
>
> But when at length he had quelled it, after having smitten it with blows, the monster fell down, lamed, and huge Earth groaned. But the flame from the lightning blasted monster flashed forth in the mountain hollows, hidden and rugged, when he was stricken, and much was the vast earth burnt and melted by the boundless vapor, like as pewter . . . So, I wot, was earth melted in the glare of burning fire. (*Theogony.*)

Is it any wonder that Velikovsky reached the conclusions he did after examining the ancient records from a catastrophist point of view? What is more to wonder is why so few have seen the truth recorded there.

Dr. Velikovsky observed:

> The scientific community starts its annals with Newton, paying some homage to Copernicus, Kepler and Galileo, unaware that the great ones of the sixteenth and seventeenth centuries searched through classical authors of antiquity for their great discoveries. Did not Copernicus strike out the name of Aristarchus of Samos from the introduction to *De Revolutionibus* before he signed imprimatur on his work? Did not Tycho Brahe find the compromising theory of the Sun revolving around the Earth—but Mercury and Venus circling around the Sun—in

> Heracleides of Pontus, yet announce it as his own? Did not Galileo read of the equal velocity of heavy and light falling bodies in Lucretius; did not Newton read in Plutarch of the Moon removed from the Earth by fifty-six terrestrial radii and impelled by gravitation to circle around the Earth, the basic postulate of Newton's *Principia;* and did not Halley read in Pliny about comets returning on their orbits? Then why does modern science disregard the persistent reports of events witnessed and recorded in many languages in the writings of the ancients and also transmitted from generation to generation by communities unable to write, by American Indians, by the people of Lapland, the Voguls of Siberia, the aborigines of tropical Africa, the Tahitians in the South Pacific? (*Velikovsky and Establishment Science,* p. 6.)

Velikovsky's book, *Worlds in Collision,* contains an excellent analysis of the evidence from ancient sources. May I highly recommend that the serious student of this subject read his writings.

"In the beginning God created the heaven and the earth. . . . Thus the heavens and the earth were finished, and all the host of them."

Genesis 1:1, 2:1

8

The New Catastrophism

Because of the recent change, historically speaking, from principles of catastrophe to principles of uniformity, men today are ignorant of the truths of the past. As Peter said, speaking of our perception of past catastrophic changes, "For this they willingly are ignorant of . . ." (II Peter 3:5.) Perhaps Isaiah best described the present condition of modern man when he said, "For the Lord hath poured out upon you the spirit of deep sleep, and hath closed your eyes: the prophets and your rulers, the seers hath he covered. And the vision of all [the prophets] is become unto you as the words of a book that is sealed . . ." (Isaiah 29:10, 11.) Because we are not schooled in catastrophism, the Bible is like a sealed book to us—we misunderstand so much.

In addition, Peter's statement that this kind of mistaken thinking would occur in the last days (II Peter 3:3, 4) strongly supports the conclusion that he was speaking of uniformitarianism, because the theory of uniformity came into being less than 200 years ago. With his visionary foresight, Peter was able to warn us against the principles of uniformity which would emerge in the last days. He made it clear that this distorted thinking would lead to ignorance and misunderstanding.

Modern science would have us disbelieve principles of catastrophism espoused by our predecessors because our predecessors were not as knowledgeable as we. However, we must remember that those men had access to information that we have lost or ignored.

Concerning the importance of understanding the scriptures from a correct point of view, Donald W. Patten, in an astute analysis of our present situation, may have put it best when he wrote:

> Until this catastrophic view of history is grasped, modern man cannot accurately conceive the perplexities, philosophies, and problems of ancient days. Once he grasps this perspective, he can begin to understand the enormous magnitude of the forces which sculptured our sphere . . . . Then only will man be able to realize how fragile his tiny planet is and how awesome is the power of his Creator. (*The Biblical Flood and the Ice Epoch,* p. 6.)

So, it seems catastrophism also becomes an attempt to reconnect man with his past. Ancient global catastrophes are an intimate part of man's heritage. His understanding of the universe around him must eventually include those precepts. That being the case, many must also acknowledge the probability of future global catastrophes, no matter how uncomfortable the thought. The prophets of God have warned that it will certainly happen again: ". . . but the heavens and the earth, which are now, by the same word are kept in store, reserved, unto fire against the day of judgment and perdition of ungodly men." (II Peter 3:7.)

Catastrophism—ancient and modern

The nature of catastrophic explanations has changed over the centuries, so perhaps it would be useful to draw a careful distinction between modern catastrophism and ancient catastrophism.

Ancient catastrophism was more a collection of subjective observations than a science. Written accounts of observed phenomena were often regarded as sacred records and therefore subjected to little critical analysis. Also, because the true nature of the solar system was generally unknown, the observed phenomena were perceived to be something other than natural events. Rather, they were seen to be supernatural occurrences, and as such were embellished by the personal and cultural biases of the chroniclers. All this fell far short of scientific observation and analysis, but it was based on observed fact—however distorted or colored those observations may have been.

Very few men throughout history correctly assessed the nature of past catastrophes. By and large, most men misunderstood the accounts handed down from generation to generation. By the time of the Dark Ages, so much superstition and dogma had been attached to ancient records—especially the scriptures—that the emerging sciences entirely rejected those records as having any validity. A few valiant souls continued to maintain, mostly on the basis of religious conviction, the validity of catastrophism; but they attempted to do so without understanding the true nature of past catastrophes. Their work, understandably, met with little or no success.

Perhaps it was inevitable that someone like Dr. Velikovsky, with keen insight and perception and unfettered by preconceived biases, would come to the conclusion that the principles of catastrophism were correct, and re-examine them in the light of current knowledge. So, modern catastrophism is also a method of analyzing ancient records and re-interpreting them in the current knowledge of the laws and order of the universe.

Modern catastrophism also includes a re-thinking of the currently accepted concepts concerning the nature of the solar system. It fixes approximate dates of ancient catastrophes, thereby giving several points of reference for reconstructing chronologies of ancient cultures. Modern catastrophism, therefore, necessitates a much more unified effort among the various scientific disciplines. (An archaeologist, for example, may also need a working knowledge of astronomy and geology in order to interpret properly what he finds in his own field.)

More and more evidence

Although catastrophism is far from being wholeheartedly accepted, each year finds more members of the scientific community leaning toward it. Increasingly, specialists in various disciplines are testing the assumptions of catastrophism in their own particular fields. The results seem to be nearly unanimous: catastrophism, with its interdisciplinary approach, is surprisingly workable and accurate in its predictions.

For example, in 1950 Velikovsky said that during past catastrophies, hydrocarbon-based flammable gases and liquids had entered Earth's atmosphere, ignited and burned over large areas of both land and water. At the time, astronomers and geologists flatly denied that hydrocarbons existed anywhere else in the solar system except on Earth. They insisted that hydrocarbons were the by-products of life, and because life existed only on Earth, there could be no hydrocarbons elsewhere.

Then science entered the space age, and man's concept of his world and the solar system began to change radically. More than thirty years after Velikovsky's prediction, the Voyager spacecraft confirmed his assumption. Voyager discovered that the Saturnian moon, Titan—which is large enough to qualify as a planet—possesses an atmosphere composed almost entirely of methane, a hydrocarbon-based flammable element! Significantly, not only is Titan's atmosphere composed primarily of methane gas, but the Saturnian moon may have great seas or oceans of super-cooled, liquid methane, together with other, more complex hydrocarbons.

If a body such as Titan were dislodged from its normal orbit and thereby fated to swing perilously close to Earth (so that the atmospheres of the two mingled), vast amounts of flammable gases would pour into our atmosphere. Igniting, they would settle onto the Earth's surface, doing incredible damage to all nature. This is only one example among hundreds that seemingly validates Velikovsky's assumptions.

Another recent theory that clearly demonstrates catastrophism is the "big bang" theory of how the universe was created. It holds that the universe was born in an explosion of incredible proportions. Though this theory, too, is meeting with some resistance from uniformitarian science, the evidence supporting it is quite conclusive. It appears as though our entire universe with its billions of galaxies came into being in a cataclysmic paroxysm of incalculable dimensions.

The most recent theory to challenge uniformitarianism was proposed by a group of scientists from Berkeley, who claim that the age of the dinosaurs was brought to its demise when the Earth was struck by a giant meteorite or comet. Such a premise would have been scoffed at thirty years ago, but it comes with much credible evidence to support its claims.

If the theory of uniformity is incorrect, as man gains more and more knowledge of the universe, it will become increasingly difficult to explain newly-discovered phenomena with uniformitarian concepts. This is exactly what appears to be happening. *The theory of uniformity is being challenged from all sides.* Recent discoveries in the fields of genetics and molecular biology have cast considerable doubt on the current theory of evolution. Space probes have demonstrated evidence of phenomena in the solar system that are unequivocally catastrophic. Recent discoveries in geology have revealed catastrophic changes in Earth's history. And so the trend continues.

Perhaps the time has come for science to cast aside its preconceived notions of uniformity and test the axioms of catastrophism.

Dr. Velikovsky's time machine

At this point, it may be helpful to mention the contribution Dr. Velikovsky has made to catastrophism and the effect his concepts have had.

From time to time, even while uniformity was the most accepted theory of change, attempts were made to show that catastrophism was a valid theory. Most of those who espoused catastrophism sought only to affirm the validity of the scriptures. Their primary concern was the question of the Deluge. However, only one or two pointed to an astral body as the cause of the Flood. Few ever considered other cataclysmic phenomena described in the scriptures to be linked with extraterrestrial agents.

Velikovsky was apparently the first person in modern times to connect accounts of ancient cataclysms with changes in the order of the solar system. Of course, science interpreted his theories as a renewal of the old dogma it had struggled so long and hard to suppress, so Velikovsky was dismissed as a reactionary and his theories as heresies, while every effort was made to suppress the publication and dissemination of his theories. A few scientists felt that Velikovsky's concepts deserved a fair hearing, but they were a small minority. Therefore, modern catastrophism had to win credibility at a slow, arduous pace.

Nevertheless, Velikovsky made many predictions concerning the nature of the planets and the solar system based on his understanding of past events as recorded in ancient literature. Most of his assertions flew in the face of then-current scientific thinking.

As the world entered the space age, however, much of what was thought to be scientific fact concerning the planets was found to be fallacy. On the other hand, the new discoveries have tended to vindicate Velikovsky and support his assertions. This has created growing interest within the scientific community, and Velikovsky's theories are being examined and explored by an ever-increasing number of scholars and scientists from many disciplines. Sadly, most of these later scientists have advanced these theories as their own, without giving recognition to Velikovsky's contributions or preeminence (though that may be, in part, due to a wish to avoid the stigma attached to Velikovsky by the majority of the scientific community).

In summary, Dr. Velikovsky bridged the gap between ancient and modern catastrophism. The theory of uniformity had caused modern man to reject the rich, recorded history of the ancients. Velikovsky forged the link that was needed to re-connect the past to the present. He pointed out that catastrophism more nearly fit the phenomena being discovered by modern science and that it more correctly explained the nature of the Earth and the entire solar system (while also confirming the evidence from ancient records of Earth's past). In addition, he correctly pointed out that the agents of catastrophic change were the other planets themselves. He honestly and forthrightly answered his critics, asking only that his theories be properly tested, rather than ignored or scoffed at. He challenged modern science to re-examine its most basic tenets in order to allow man to correctly understand creation.

"And ye shall know the truth, and the truth shall make you free."

John 8:32

9

The Bible— Right or Wrong?

One of the most difficult struggles for Christianity today is the effort to establish the Bible as accurate history. As never before, the Bible has come under attack for claims that it makes concerning the creation of the world and man, asserting "extravagant accounts" such as the Deluge and the great day of Joshua to be real events. The debate is as heated as ever.

In the midst of this ongoing turmoil, Dr. Velikovsky introduced his revolutionary concepts of Earth's history. Strangely, though his theories seemingly hold some of the answers to the aforementioned debate, little attention was given him by the Christian community. Therefore, it may be valuable to examine the theories of catastrophism as they apply to the Bible in order to see how they validate scripture and give men of reason new understanding of its profound message.

Evolution vs. Creation

Ever since Charles Darwin published his revolutionary book, *The Origin of Species by Means of Natural Selection* (1859), advancing his theory of biological evolution, an ideological battle has been fought between the Creationists and the Evolutionists. The Creationists are, for the most part, religious fundamentalists who accept the Biblical story of man's creation literally. The Evolutionists, on the other hand, affirm that a rational analysis of the flora and fauna whose remains have been found in the geologic record, leads to the conclusion that the higher forms of life—man being the highest—have evolved, over vast eons of time, from the lower forms of life. Predictably, the Creationists take offense at such statements because they believe that the first man was created *by God* from the very elements of the Earth, that random evolution had nothing to do with it.

Of course, there are some strange combinations of the two theories that are proffered from time to time in an attempt to reconcile the conflict. (For example, a believing Christian who is also a scientist may consider that evolution is the method that God used to "create" man; another may consider that all life did evolve—except for man—who was unique and different in that he was "created." Such approaches leave us divided and opinionated concerning our own origins.)

The Latter-day Saints have yet another perspective of the creation. Brigham Young, with his special insight, said this concerning the creation of man:

> Here let me state to all philosophers of every class upon the earth, when you tell me that Father Adam was made as we make adobes from the earth,

> you tell me what I deem an idle tale. When you tell me that the beasts of the field were produced in that manner, you are speaking idle words devoid of meaning. There is no such thing in all the eternities where the Gods dwell. *Mankind are here because they are the offspring of parents who were first brought here from another planet,* and power was given them to propagate their species, and they were commanded to multiply and replenish the earth. (*Journal of Discourses,* 7:285-286, italics added.)

Then according to Brigham Young, the answer to the question of creation or evolution on our planet is: neither is correct! All things that have ever existed upon this Earth were *brought here* from elsewhere. Our planet was seeded with life from another planet.

In light of Brigham Young's statement, it is amazing to learn that "Sir Fred Hoyle, the British astronomer, produced statistical evidence suggesting that many virus diseases had their origin in space." ("Was Darwin Wrong?", *Life,* April, 1982, p. 52.) If a virus can be introduced into Earth's ecosystem from space, then logic and an understanding of cataclysmic principles leads to the conclusion that higher forms could also be introduced; and if all life on Earth was brought here from elsewhere, as Brigham Young said, then the catastrophe model of worlds intersecting with one another from time to time, supplies the mechanism whereby life could feasibly be carried from one world to another. Once again the theory of catastrophe and the teachings of latter-day prophets appear to complement one another.

Darwinism discredited

The problem of natural selection is easily answered by the theory of catastrophe, even without modern revelation on the subject. An analysis of Darwinism from the point of view of catastrophism may be instructive.

Initially, it should be pointed out that Darwinism (or evolution by natural selection) is only biological uniformitarianism. Almost two centuries ago, the then-emerging geological theory of uniformity needed a companion theory in biology to lend credibility to its concept that greatly expanded periods of time made up the Earth's history. Darwin's theory did just that. It held that the slow geologic processes of sediment deposition, settling land-masses, and mountain formation were matched by an equally slow process of natural selection among life forms. Darwin theorized that the process of natural selection was caused by slowly changing environments, allowing the flora and fauna to evolve in order to meet the demands of their changing habitat.

Because evolution is only the biological counterpart of the theory of uniformity (and hence an offshoot of uniformitarianism), most modern catastrophists see evolution as a flawed theory. This is so because modern catastrophism holds that the changes in the geologic order of things have been, and will be, *sudden and quick*—so much so that geologic events for which the uniformity theory allows tens of thousands of years can be accomplished in a few years or less. Therefore, according to the theory of catastrophism, there simply would not be enough time for the process of natural selection to operate.

This should be heartening news to Latter-day Saints. Here is a theory that not only validates Biblical people, places, and events but also supports the Biblical time frame for the creation of the Earth and provides the mechanism for placing life on the Earth. However, a word of caution may be in order here. It is unlikely that the secular catastrophists will ever accept the fundamentalist's theory of creation. By the same token, Latter-day Saints see the creation from yet another point of view (as described in the aforecited quote by Brigham Young). Understanding the catastrophe model of intersecting planets adds yet more information to our knowledge of the history of our planet and the life that resides thereon.

Velikovsky himself was not a creationist—he proposed catastrophic evolution, in which species changed suddenly through multiple mutations on a massive scale, accompanied by total extinction of other species. Nonetheless, his theories discredit the uniformitarian theory of evolution which held sway during the past century. It should not be surprising to expect that as catastrophism continues to investigate the matter, the currently accepted theory of evolution will be found to be as useless a theory as the theory of uniformity that spawned it.

The Bible is accurate history

Another topic that causes friction between the scientific and religious communities is the Biblical account of the great flood, which only Noah and his family survived. Along with the Biblical concepts of creationism and catastrophism, science rejected the Deluge as well—for much the same reasons. However, modern catastrophism points to the universality of the flood tradition as an indication of its *validity*. There are abundant records, myths, and legends of a great flood. One survivor of that cataclysm is known to us as Noah. The Greeks called him Noeus, the Babylonians called him Ut-napishtim, and the Aztecs called him Coxcox (the "x" has an "sh" sound).

The Deluge is a most difficult topic (as is any discussion of the times prior to it) because it belongs to man's most distant past—hoary antiquity. Even the Bible has little to say about those times. It took modern revelation to give us some additional insight into the events of mankind's earliest days on this planet. So little evidence exists in written records, that it is nearly impossible to reconstruct the events of the Flood. But the evidence that does exist clearly indicates that it was *an incredible catastrophe*, and catastrophists unanimously point to an extraterrestrial agent and changes in the solar system as the cause.

Again, modern catastrophism seems to hold the answers for Christians in answering the critics of the scriptures.

Joshua and Jericho

Another event recorded in scripture that is scoffed at by scientists is the story of Joshua, the great General who succeeded Moses as the leader of the Nation of Israel. Joshua led his people in their conquest of the Promised Land. His first triumph is known in history as the Battle of Jericho.

Modern archaeological excavations have confirmed that ancient Jericho was protected by a tremendous wall and that it was destroyed shortly after the end of the Egyptian Middle Kingdom (approximately 1450 B.C.) by a great earthshock. But because the Exodus was thought to have occurred during the Late Egyptian Kingdom, archaeologists believed that the Israelites arrived at Jericho long after the great wall had fallen—several hundred years later, in fact. Hence, many expressed the sentiment that the Old Testament, once again, had been proven false—that the story of the Battle of Jericho was another Hebrew myth. That's where things have stood for several decades.

Velikovsky, with his revolutionary reconstruction of Egyptian chronology, set the record straight. The evidence was confusing because the time fixed by scholars for the Exodus was incorrect. If, as Velikovsky suggested, the Exodus occurred at the end of the Egyptian Middle Kingdom rather than hundreds of years later, the Israelites would have arrived at Jericho precisely at the time archaeologists indicated for the collapse of the great wall.

Velikovsky concluded that the Exodus had occurred at the end of the Middle Kingdom, because of the *Papyrus Ipuwer*. That record, written by an Egyptian scribe, had been translated and dated to the end of the Middle Kingdom by Sir Alan Gardner—not by Dr. Velikovsky. Therefore, the events described by Ipuwer (fixing the Exodus at the end of the Middle Kingdom) and evidence from excavations at Jericho (fixing the collapse of its great wall shortly after the end of the Middle Kingdom) properly coincide. Velikovsky's reconstruction of ancient history answered this riddle and vindicated the Old Testament account of the Battle of Jericho.

The day the Earth stood still

Even more remarkable is Velikovsky's verification of the second great event in Joshua's campaign to conquer the promised land—the day the Earth stood still!

Of all the Old Testament stories, Joshua's account of the sun and moon standing still has been singled out by scientists as the most laughable. Yet, as was pointed out earlier, Velikovsky maintained that the events of that day, as Joshua described them, marked the return of the great comet/planet of the Exodus.

Fifty years earlier (immediately after the events of the Exodus), the Earth and the comet Venus had parted ways, each pursuing its own path in space. However, that first encounter had somewhat altered the orbits of the two worlds so that they would again coincide approximately every fifty years. (The return of the comet/planet Venus every fifty years may account for the Hebrew tradition of the Jubilee year, and the fifty-year century of other ancient cultures, including the Meso-American Indians.)

The second encounter between the Earth and the comet/

protoplanet occurred as Joshua pursued the Amorites in battle:

> And it came to pass, as they fled from before Israel, and were in the going down to Beth-horon, that the Lord cast down great stones from heaven upon them unto Azekah, and they died: they were more which died with hailstones than they whom the children of Israel slew with the sword. (Joshua 10:11.)

Notice the preliminary event: "great stones" which fell and killed the Amorites—a clear indication as to the nature of the cosmic event to follow:

> And the sun stood still, and the moon stayed. . .
>
> So the sun stood still in the midst of heaven, and hasted not to go down about a whole day. (Joshua 10:13.)

The book of Joshua does not bother to list all the manifestations that occurred at that time, but rather it refers to another book: "Is not this written in the Book of Jasher?" (Joshua 10:13.) This book of Jasher is part of a large group of ancient documents that, for one reason or another, were not included in the Bible. In it we find more evidence of the signs we might expect to see if this event was cataclysmic in nature. Joshua praises the Lord for this deliverance, saying:

> Thou hast done mighty things, O Lord, thou hast performed great deeds; who is like unto thee? . . .
>
> The sun and moon stood still in heaven, and thou didst stand in thy wrath against our oppressors and didst command thy judgments over them.
>
> All the princes of the earth stood up, the kings of the nations had gathered themselves together,

> they were not moved at thy presence, they desired thy battles.
>
> Thou didst rise against them in thine anger, and didst bring down thy wrath upon them; thou didst destroy them in thine anger . . .
>
> Nations have been consumed with thy fury, kingdoms have declined because of thy wrath, thou didst wound kings in the day of thine anger . . . .
>
> . . . thou didst terrify them in thy wrath, and didst destroy them in thine anger.
>
> The earth trembled and shook at the sound of thy storm over them . . .
>
> Thou didst persue them in thy storm, thou didst consume them in thy whirlwind, thou didst turn their rain into hail, they fell in deep pits so that they could not rise.
>
> Their carcasses were like rubbish cast out in the middle of the streets. (*The Book of Jasher,* pp. 247, 248.)

In this text (which amplifies on the laconic Old Testament account), the sure signs of a cosmic catastrophe can be seen. (For an elaboration of the signs of a cosmic catastrophe see *And The Moon Shall Turn To Blood,* Chapter 3.)

Velikovsky cited accounts in the records from other ancient cultures of a time when the inhabitants saw either a prolonged day (as did the Israelites) or a prolonged night (as did the Meso-American peoples), depending upon where they were located on the Earth. (See *Worlds in Collision,* pp. 55-62.) The extra-Biblical sources he cited are numerous and emphatic. The world was again afflicted by the great comet, only to a somewhat lesser degree than in the first encounter. The long day of Joshua was not myth or folklore; it was a historic fact. (See also *And The Moon Shall Turn To Blood,* Chapter 5, for a discussion of the correlation of the events of the Long Day of Joshua with statements by Nephi.)

It appears that catastrophism lends credibility to the Biblical accounts of the Flood and the Long Day of Joshua. Velikovsky's explanation of history clearly vindicates Christians for tenaciously holding that the Bible was correct history. These few examples form only a small part of the mounting evidence which demonstrates that the Bible is correct.

"But of that day, and hour, no one knoweth; no, not the angels of God in heaven, but my Father only."
Joseph Smith—Matthew 1:40

10

When Will It Happen?

Once the concept of ancient catastrophes is introduced and its relevance to prophecy explained, the reality of imminent cataclysmic events compels the inquirer to ask, "When will it happen?" Once the prophecies are made more plain, it seems natural to want to know "when" these events will take place. Beware of this path.

Many try to wrest from the scriptures the day or the year these events might occur. It is popular among modern Christian ministers, evangelists, and millennialists to speculate endlessly about when the Savior might return, because they believe they see the fulfillment of Biblical prophecy in this present day and time. In so believing, they are correct (but rather tardily) since the Lord revealed to Joseph Smith that these were the last days as early as 1820. That has only become evident to the rest of Christianity as they have seen some prophecy begin to be fulfilled.

Those who search for the key to "when" primarily quote the books of Daniel and Revelation in an effort to make sense of the periods of time recorded in association with events of the last days. Even Latter-day Saints have been caught up in this endeavor, though the Lord has said "of that day and hour knoweth no man." (D&C 49:7.)

The dates given in Daniel and Revelation have been interpreted at various and sundry times down through the ages with no apparent success. In each case the person or persons doing the interpreting have "discovered" that the date applied to their own day and time. There are records of people in the 1200s, the 1500s, the 1800s, and now in the 1900s interpreting the dates in prophecy to apply to their particular day and time. Obviously none were correct, since the Second Coming has not yet occurred.

Misled seekers

Many are the accounts told of men and women who foolishly assembled to meet the Savior on the "prophesied" day, only to be dismayed to see nothing out of the ordinary occur. Such as the fate of the Seventh-day Adventist movement in the early 1800s. William Miller (co-founder of the Adventist movement) began predicting as early as 1831 that the Second Coming would occur in the year 1843. (He was a contemporary of the Prophet Joseph Smith, and so as might be expected, the Prophet had something to say concerning the predictions of Miller.) The Prophet's assessment of the situation clearly pointed out Miller's error before the fact:

> I showed them the fallacy of Mr. Miller's data concerning the coming of Christ and the end of the world, or as it is commonly called, Millerism, and preached them quite a sermon; *that error was in the Bible; that Miller was in want of correct information upon the subject,* and that he was not so much to blame as the translators. I told them the prophecies must all be fulfilled; the sun must be darkened and the moon

> turned into blood, and many more things take place before Christ would come. (*Documentary History of the Church,* Vol. 5, p. 272.)

According to the prophet, the error of Miller and those who followed him was to focus on the dates of the prophesied events, while ignoring the prophesied signs. Two months after this statement was made, on the day fixed by Miller for the Savior's advent, the Prophet made this note in passing: "Monday, April 3, [1843],—Miller's day of judgement has arrived, but it is too pleasant for false prophets." (Ibid., p. 326.)

Unreliable dating

If there is one lesson to be learned with respect to prophecy, it is that no interpretation of the time periods specified can be considered reliable without also considering the prophesied signs as well. In addition, the time periods specified in prophecy are not given that we may calculate the day or year of a given event. W. Cleon Skousen made note of that fact:

> . . . the Lord rarely, if ever, signifies the exact date when a prediction will come to pass. He will disclose the period and the accompanying circumstances but not the precise date. (*Prophecy and Modern Times,* pp. 4-5.)

If the calculated date for a prophesied event approaches—the Second Coming, for example—and yet the signs prophesied to accompany that event are not emphatically manifest, then the logical conclusion would be that the interpretation of the timing was wrong. In addition, as the Prophet pointed out in the above quotation, certainly the dates in scripture have been subject to error and misinterpretation by copiers and translators. More than any other aspect of prophetic pronouncements, the numbers and dates associated with future events are among the most susceptible to error and misinterpretation.

What serves to make the task even more difficult is the fact that there are very few places in the scriptures where the duration of future *periods of time* are given. Since there exists such a dearth of information on the subject, it becomes virtually impossible to correlate the facts and thereby substantiate them to any degree. Such correlation is imperative if sense is to be made of the information. The *signs*, on the other hand, are found everywhere in the scriptures. They are enumerated over and over again as if to insure that the reader will understand them. Such repetition of the signs allows careful correlation and substantiation. Hence, a study of the "times" may be fruitless and futile, while a study of the "signs" appears to be most profitable.

Seeing that so many attempts at deciphering the date of the Second Coming have failed and considering Joseph Smith's negative assessment of such efforts, perhaps it would be well to refrain from attempting to do so. It would seem, as the Prophet implied, that the prophesied signs would be much more reliable indicators of the proximity of the Second Coming. There is an abundance of information available concerning the signs of the last days in scripture, so they can be cross-referenced and compared to give a rather complete picture of events leading up to the Savior's appearance.

In addition, the Lord himself has said:

> *Search these commandments,* for they are true and faithful, and the prophecies and promises which are in them shall all be fulfilled.
>
> What I the Lord have spoken, I have spoken, and I excuse not myself; and though the heavens and earth pass away, my word shall not pass away,

> but shall all be fulfilled, . . . (*Doctrine & Covenants* 1:37-38, italics added.)

Learn the signs

The abundance of prophecy in the scriptures, touching on latter-day events, is ample evidence of God's desire that we understand the upcoming events so that we might not prematurely anticipate the Savior's appearance (Joseph Smith —Matthew 23, 37, 39-40), and so that we might be emotionally prepared for the overwhelming devastation of the destructions to come. "And all things shall be in commotion; and surely, men's hearts shall fail them; for fear shall come upon all people." (*Doctrine & Covenants* 88:91.)

Perhaps the intent of the Lord in revealing so much of the signs is so that we may learn the order, magnitude, and nature of future events. It may be that knowing the exact date of the Second Coming would, in effect, take away our free agency. The effort to learn the year or the hour when the Savior will appear is a futile and fruitless one. The Savior used a parable to explain this:

> But know this, if the good man of the house had known in what watch the thief would come, he would have watched, and would not have suffered his house to have been broken up, but would have been ready. Therefore be ye also ready, for in such an hour as ye think not, the Son of Man cometh. (Joseph Smith—Matthew 47-48.)

Clearly, the "good man" will not know when his house will be "broken up." So it appears that we are purposely left ignorant of the "hour" of His coming in order to learn whether we will ignore the warnings or whether we will repent, prepare, and be ready at all times.

The purpose of prophecy

Skousen noted the purpose of prophecy:

> The purpose of prophecy is frequently misunderstood. It is not the Lord's purpose to completely remove the veil which guards the future. Too much prophecy would defeat the plan of salvation. Life on earth was designed for our education and testing . . . .
>
> To reveal too much would inevitably overwhelm the free agency of man and deprive us of our chance to choose. (*Prophecy and Modern Times*, pp. 5-6.)

Perhaps the best example of what prophecy can tell us lies in the prophesies of the return of the Jews to Palestine and the creation of the state of Israel in modern times. The most basic premise of all the millennialist writings today deals with prophesies of the Second Coming as it relates to the gathering of Israel. It is affirmed in prophecy (by prophets both ancient and modern) that the Jews will be gathered to their homeland prior to the paramount event of the last days—the Second Coming. The prophets testified again and again that Judah would be scattered among the Gentile nations, only to be gathered again to their homeland in the last days before the Messiah would come.

Jerusalem was destroyed by the Romans in 70 A.D. At that time the Jews were exiled from Palestine and scattered throughout Eurasia. For centuries they remained scattered among the nations of the world. Then in 1841, the Lord, through Joseph Smith, sent Orson Hyde to Jerusalem, and in an inspired prayer spoken on the Mount of Olives, he dedicated the land to the return of the House of Israel. Subsequently, in the late 1800s the Zionist movement began, and Jews began returning to their homeland in ever increasing numbers.

Because the Jews began returning to the Palestine area in significant numbers and because a new state of Israel was carved out of the Middle East in the late 1940s, all students of prophecy naturally expect the Second Coming to occur at any time. To all appearances, the part of the prophecy predicting the return of the Jews to their homeland has been fulfilled in this generation before our very eyes! The prophets predicted this order—this particular sequence of events: first the return of the Jews to Jerusalem, then the Second Coming. Therefore, we may expect to see the prophesied events transpire in the predicted order.

No man independent of revelation could predict exactly when the first part of the prophecy—the gathering of Judah—would be *fulfilled.* Even the Prophet Joseph Smith in his time did not attempt to fix a time, except to say that their gathering was about to begin. In a like manner, no uninspired man today can say when the second part of the prophecy—the Second Coming—will occur, even though prophetic events seem to indicate its imminence.

The prophet will know

Though we may not know for ourselves when the Savior will come, He will certainly inform the prophets. "Surely the Lord God will do nothing, but he revealeth his secret unto his servants the prophets." (Amos 3:7.) The Book of Mormon contains a fine example of this. Nephi (grandson of Helaman, in the meridian of time) appealed to God that the righteous might not be slain by the wicked, who had determined to rid themselves of the foolish believers because the signs of the Savior's first coming had not yet appeared as prophesied:

> And it came to pass that he cried mightily unto the Lord all that day; and behold, the voice of

the Lord came unto him, saying:

> Lift up your head and be of good cheer; for behold the time is at hand, and on this night shall the sign be given, and on the morrow come I into the world, to show unto the world that I will fulfil all that which I have caused to be spoken by the mouth of my holy prophets. (3 Nephi 1:12-13.)

Much the same thing could be said about the fulfilling of prophetic events in our day. *The Lord will reveal the impending events to the prophets.* The prophets, in turn, will direct the righteous to prepare in the proper manner. (Indeed, some might say that they have already done so.) There is ample precedent for this conclusion in the prophetic leadership of men like Jared, Moses, Noah, Abraham, and many others. In each case, God revealed to these men what they, and the righteous who would follow them, could do to avoid injury or death while God poured out His judgments on the world.

Seeking to discover the date of the Savior's coming is folly, at best. We should search the scriptures to understand the nature of upcoming events and heed the direction of the Prophet. He will know, and through him we may know. The lesson of history instructs us to *search the scriptures to know how to live and what to expect in the future, and to look to the living prophet for guidance and direction.*

". . . I will rend their kingdoms; I will not only shake the earth, but the starry heavens shall tremble. For I, the Lord, have put forth my hand to exert the powers of heaven; ye cannot see it now, yet a little while and ye shall see it . . ."

Doctrine & Covenants 84:118, 119

11

Where Will It Come From?

The question of *where* the "intruder" of the latter-days will come from is a most interesting one, since knowing what to look for may aid us greatly in our preparations for the coming events.

Past events indicate that we will see it coming if we know what to look for. Velikovsky indicated that anciently all the world witnessed the great "signs" in the heavens, prior to being afflicted by them. He pointed out that though the signs tooks many forms, they (and the destructions that followed them) could all be traced to a single source—errant heavenly bodies.

The New Testament and the Book of Mormon express much the same thing. They speak of the appearance of a "star" and "signs" in heaven at the Savior's First Coming—his birth.

(Matthew 2:2, 3 Nephi 1:21.) In addition to the new star, according to the Nephite account, there was another sign—a night without darkness. Both records speak of destructions in conjunction with His death (the Western Hemisphere having suffered far worse destruction than the Eastern). It may be worthy of note, however, to observe that the signs of destruction preceded His appearance among the Nephites, just as they will in the last days. Clearly, something extraordinary was occurring in the heavens during both the Savior's birth and death. The Savior's first appearance on Earth seems to have set a precedent: signs would appear in heaven and on the Earth preceding His coming, whether it was the first or second.

But what should we look for, and where will it come from? Though there may be other possibilities, to date this author is aware of only three.

From the edge of the solar system

The first possibility is for a large comet to enter the known solar system from beyond the orbit of Pluto (from an area known as the "cometary halo") and perturb one or more of the known planets (thus upsetting the order that presently prevails) and setting the stage for a series of near collisions.

There are an undertermined number of bodies in the cometary halo—some estimate as many as a million—that are orbiting our Sun. There is much speculation about the size of such bodies. Undoubtedly they vary widely. Some may be the size of an asteroid—varying from a few feet to tens of miles in diameter; still others may be as large as a moon; and some even as large as a planet.

All such bodies follow elongated eliptical orbits which carry them far out into interstellar space at their *aphelion* (the point at which they are furthest from the Sun), and they seldom approach closer to the Sun than the orbits of Jupiter or Saturn at their *perihelion* (the point at which they are closest to the Sun). Because they do not enter the inner part of the solar system, they do not grow the familiar tail and thus go undetected by all but an occasional astronomer.

These bodies are infrequent visitors to the known solar system because of the length of their orbits. As a point of comparison, the most distant known planet is Pluto. It takes Pluto 248 plus Earth years to complete one orbit. Any planetary body out beyond the orbit of Pluto would have an even longer orbit. Some bodies travel so far out and away from our sun before returning that they likely only visit the known solar system once every few thousand years. Hence it is possible, as Velikovsky speculated, that

> A large comet arriving from interstellar spaces may run into one of the planets and push it from its orbit; then chaos may start anew. Also, some dark star, like Jupiter or Saturn, may be in the path of the sun, and may be attracted to the system and cause havoc in it. (*Worlds In Collision*, p. 375.)

From deep space

The second possibility again looks to space, far beyond our solar system. It may be that an object known to astronomers as an "ultramassive body" would pass near or through our solar system, distorting the orbits of one or more planets with its incredible gravitational pull, and thus set the stage for renewed collisions.

Michael Simon Bodner (a physicist) and Michael E. Brandt (a graduate biomedical engineering student) "speculated about the possibility of a very massive object such as a neutron star, black hole, or white dwarf passing through or near the solar system." They concluded that such a body, commanding tremendous gravitational power (far exceeding its relatively small size and moving through space at speeds amounting to an appreciable fraction of the speed of light) would have a profound effect upon our solar system. They further noted that many unexplained conditions in the solar system could be explained by such a thesis: the "tipped" axis of Uranus which causes it to appear to roll along in space much like a barrel; the elliptical orbit of Pluto which takes it inside the orbit of the next closest planet to the sun, Neptune; Earth's capture of its moon; the asteroid belt between Earth and Mars (likely the remains of a shattered planet); rings around Saturn; and other strange phenomena. Such evidence of past catastrophe and disorder in the solar system tells of previous cataclysmic disruption. ("The Gravitational-electromagnetic Effects of Ultramassive Objects On The Solar System," *Kronos*, Vol. VI, No. 2, pp. 3-17.)

Such a body as described above could pass through our solar system in only a few minutes, giving precious little advance notice of its approach. In fact, there may or may not be a warning. If it were a light-emitting body, it would be spectacular. If it were a dark body, it probably would not be seen approaching.

However, if it were a dark body, a logical scenario can be constructed wherein we would have a warning of sorts. For example, if such a massive body were created in a super-nova (star explosion) that was, say, ten light years removed from Earth, and this ultramassive object were moving at one-third the speed of light toward our solar system, then the following sequence of events would be observed from Earth: we would see a bright new star born (the super-nova); it would shine brightly for a time, then dim; thirty years later the massive object created in the super-nova would crash through our solar system at nearly 100,000 miles per second; passing some distance away from Earth and moving rapidly, it's influence would be felt for up to two or three hours—directly affecting only one hemisphere of Earth by its massive gravitational pull. If it passed close enough to have a direct effect upon the Earth without totally dislocating it from its orbit, it would certainly cause a terrible disruption of nature on Earth; and finally, earthquakes, hurricanes, tides, volcanic ash, and other phenomena on Earth would be the result. Note how the foregoing scenario matches the description of events at the time of Christ's birth and death. It also bears a striking resemblance to the calamity described in Revelation, after the opening of the sixth seal. (Remember, this comes from two scientists who know nothing of the Book of Mormon account which this so nearly parallels. They reached this conclusion in the process of following up on Velikovsky's concepts.)

At some time in the future, the prophesied events may be initiated suddenly in the chaos created by an ultramassive object passing through our solar system. Such an intruder would undoubtedly alter the orbit of one or more planets, resulting in intersecting orbits. Intersecting orbits would, in turn, result in a new series of catastrophic events.

From a Jovian planet

The third possibility deals with the same mechanism by which Venus—indeed all the inner planets—may have been created: by splitting off from the core of one of the gaseous giants. In 1977 Eric W. Crew, a Fellow of the Royal Astronomical Society, wrote:

> The suggestion that the solid cores of gaseous planets may be ejected periodically, sometimes achieving independent orbits about the parent star, may solve a very important, long-standing problem in Solar System astronomy: that of the origin of the dense, "terrestrial" planets and satellites. The large

> gaseous planets would have the volumes and masses to attract and retain enough heavy cosmic materials to form large cores in much shorter times than would be required for the same masses of solid material to accumulate without this assistance. ("Stability of Solid Cores in Gaseous Planets," *Kronos*, Vol. III, No. 1, pp. 18-26.)

Recall that Velikovsky said that Venus was originally ejected from the planet Jupiter. Crew, in the above article, speculates that the giant Red Spot in Jupiter's atmosphere may be the site where Venus emerged from the hot, dense core. Hence, if Velikovsky's assumptions are correct, then the chaos predicted in prophecy may be initiated in a very similar manner to that which occurred anciently: a new planet will be ejected into the solar system from within one of the gaseous giants, and it would careen across the heavens colliding or nearly colliding with the inner planets.

A likely scenario

From the statements of John in Revelation, it appears that the future will see a combination of the aforementioned possibilities. At the opening of the sixth seal, John wrote:

> . . . and, lo, there was a great earthquake; and the sun became black as sackcloth of hair, and the moon became as blood; and the stars of heaven fell unto the earth . . . . And the heaven departed as a scroll when it is rolled together; and every mountain and island were moved out of their places. (Revelation 6:12-14.)

These are familiar signs, seen when worlds nearly collide. But a close inspection of the individual elements of this catastrophe reveals certain peculiarities. It appears that during this particular event there will be no bloodied water, no meteorites falling from the skies, no burning petroleum or flammable gases. All these missing elements point to one interesting fact: there appears to be no sharing of atmospheres between Earth and the body which causes this calamity. A comet-like body with its attendant tail/atmosphere would certainly deposit some of its debris in Earth's atmosphere, just as did the comet of the Exodus. Each of the "woes" that are enumerated later in Revelation are accompanied by signs of such debris descending upon the Earth—bloody water, hail and fire. This first event, then, is clearly dissimilar to the "woes" that follow it.

Could it be that what John described in this event was the result of an ultramassive body passing through our solar system? Since the above quote from Revelation omits any reference to material descending into Earth's atmosphere, it seems logical to conclude that the passing body will have no atmosphere. This would be more typical of an ultramassive body than a cometary body. If so, then it seems likely that the first event described by John in the sixth chapter of Revelation will happen suddenly, with very little warning. Of all the possibilities discussed in this chapter, the ultramassive body theory is the least likely to give a clear warning. That fact, too, fits with the description of the initial catastrophe listed in Revelation. Further, the intruding body would disrupt the order that currently exists among the planets, causing them to intersect as they did anciently. John's description of three "woes" to follow the initial calamity also fits this pattern of subsequent near collisions among the disrupted planets.

Of course, *this is strictly speculation.* However, there is one thing that prophecy makes plain: the world will be menaced once again by *bodies from space,* just as it has been in the past. Prophecy clearly declares it, and the prophecies will be fulfilled. Whether we see it coming or not is of little consequence, but it is interesting to attempt to understand the mechanisms whereby God orders the Universe. For if we are to know about God, as with any individual, then we must study His works.

"And there shall be men standing in that generation, that shall not pass until they shall see an overflowing scourge; for a desolating sickness shall cover the land."
Doctrine & Covenants 45:31

12

The Great Plague

Though the following descriptions from prophecy are rather graphic and unpleasant to ponder, it reveals useful information about what to expect and how to prepare beforehand.

The first sign of the Exodus was the dust that settled to earth on land and water, turning everything red—especially the water. In addition, something like "ashes of a furnace" fell on Egypt. All this alien debris polluted open water sources, such as lakes and rivers, so that no one could drink the water. It was also fatal to marine life. Prophecy tells of a time when rivers, seas, and oceans will be affected by "a great star from heaven" which will pollute the waters of the Earth:

> And the name of the star is called Wormwood: and the third part of the waters became wormwood; and many men died of the waters, because they were made bitter. (Revelation 8:11.)

Alien pollution

In his remarkable vision of the future, LDS Patriarch Charles Evans in 1894 wrote this about the red, polluted water:

> I saw in my dream the messenger again appear with a vial in his right hand, who addressing me, said: 'Thou knowest somewhat of the chemistry taught in the schools of human learning, behold now a chemistry sufficiently powerful to change the waters of the sea.'
>
> He then poured out his vial upon the sea and it became putrid as the blood of a dead man, and every living soul therein died. (*The Contributor,* [an LDS publication], Vol. 15, No. 20, p. 642.)

John, in Revelation, also wrote "the second angel poured out his vial upon the sea; and it became as the blood of a dead man." (Revelation 16:3.) Wilford Woodruff recorded the same thing seen in vision, only the condition of the water was attributed to "the putrification of dead bodies." He recorded:

> The waters of the Chesapeake River and of the city were so stagnant and such a stench arose from them on account of the putrification of dead bodies that the very smell caused death. (A vision recorded in the *Wilford Woodruff Journals,* June 15, 1878.)

But because the tainted water itself will become "putrid as the blood of a dead man," it is likely that the dead bodies only contributed to its foul condition. However, President Woodruff did understand the nature of the coming cataclysm and, in particular, the effects of the great plague:

> You are about to be visited with war, the sword, famine, pestilence, plague, earthquakes, whirlwinds, tempests, and with the flame of devour-

> ing fire . . . . The seals are about to be opened; the plagues to be poured forth. *Your rivers and seas will be turned into blood and to gall.* (*Millennial Star,* 41:241, 245-6, italics added.)

Since Patriarch Evans' vision described the effect of the material on the water as chemical in nature (which turned the water "putrid as the blood of a dead man"), then it would be entirely reasonable that the waters were also polluted by putrefying bodies.

No drinking water

It is certainly an understatement to say that at such a time there will be an extreme shortage of drinking water. It appears likely that we should prepare for that eventuality.

During any localized natural calamity, drinking water is a primary concern. People suffer from thirst in the aftermath of floods and earthquakes as much as they would in a sudden drought. In future world-wide catastrophes the specter of water, poisoned by extraterrestrial material, will make the hardship even greater.

It appears, therefore, that we would also be wise to consider where we might find potable water during future events that will pollute otherwise good sources. It is unlikely that the delivery systems we now depend upon for water will be functional after the destructions predicted. The major movements of Earth's crust predicted to occur will alter the present location of our water sources. Dams and reservoirs will be destroyed. Old wells will dry up, new ones will appear in new locations.

Where, then, should we look for water? The answer is not as difficult as it appears. When open water sources were contaminated anciently, the Egyptians dug wells. Apparently, water from wells had either not yet been contaminated by fouled surface water, or the contamination had been filtered from the subterranean water as it made its way down from the surface. In either case, water from underground aquifers was the only safe water. (Worthy of note is the fact that Moses and the Israelites suffered for want of water. They searched for the naturally-occurring artesian "wells in Sinai," after leaving Egypt [Exodus 15:27, 17:6]. Open water sources in the Sinai were also polluted and bitter [Exodus 15:23].)

Dust induced illness

What is even more ominous, the dust that fell in Egypt anciently apparently irritated the skin, causing boils and blains, disease, and death. (Exodus 9:3, 9.) Again, it appears that the same thing will hold true in the future. In Revelation, John is more explicit about this first plague in the third "woe," associated with the star called Wormwood:

> And the first went, and poured out his vial upon the earth; and there fell a noisome and grievous sore upon the men which had the mark of the beast, and upon them which worshipped his image.
>
> And the second angel poured out his vial upon the sea; and it became as the blood of a dead man; and every living soul died in the sea.
>
> And the third angel poured out his vial upon the rivers and fountains of waters; and they became blood.
>
> And they gnawed their tongues for pain [thirst?], and blasphemed the God of heaven because of their pains and their sores, and repented not of their deeds. (Revelation 16:2-4, 10-11.)

Apparently, the Lord was referring to this great, worldwide plague when He said to Joseph Smith:

> And there shall be men standing in that generation, that shall not pass until they shall see an overflowing scourge; for a desolating sickness shall cover the land. (Doctrine & Covenants 45:31.)

Then death

In the vision recorded by Elder Woodruff, a disease was referred to which caused men to "roll over a time or two and die in agony." In addition, Patriarch Evans gave this remarkably detailed account of the plague associated with the pollution of Earth's atmosphere and water:

> Again the light shone, revealing an atmosphere tinged with a leaden hue, which was the precursor of an unparalleled plague whose first symptoms were recognized by a purple spot which appeared on the cheek, or on the back of the hand, and which, invariably enlarged until it spread over the entire surface of the body, producing certain death . . . . Wicked men, suffering under its writhing agonies, cursed God and died, as they stood on their feet, and the birds of prey feasted on their carcasses. (Op. cit.)

Hemorrhaging of the capillaries (the smallest blood vessels in the body) that lie just beneath the skin would account for the appearance a purple spot, and may be a clue to the nature of the plague and its effects on living organisms. Coming in contact with the skin, the compound would begin to affect the skin and the tissues immediately beneath the skin—inducing the hemorrhaging. Patriarch Evans wrote that the spot "enlarged until it spread over the entire surface of the body, producing certain death." That would indicate massive hemorrhaging within the body. If that is so, then inhaling the compound would have an even more detrimental effect on the lungs. Whereas the skin would act as a barrier (slowing the attack on the body), the capillaries and the tissues around the lungs would rupture when contacted by the compound, filling the lungs with fluid. The victims would die of asphyxiation—having drowned in their own body fluid.

Lethal compound

Patriarch Evans was told that the change in the water was a chemical one, and it seems clear from this statement from his vision that the effect of the compound on human tissue will be devastating:

> This plague, in grown persons, rotted the eyes in their sockets and consumed the tongue as would a powerful acid or an intense heat. (Op. cit.)

The Lord himself described the same effects of the plague to Joseph Smith as he enumerated the signs, plagues, and desolations that will precede the Second Coming.

> And their tongues shall be stayed that they shall not utter against me; and their flesh shall fall from off their bones, and their eyes from their sockets . . . . (Doctrine & Covenants 29:19.)

It appears from the accounts related by Elder Woodruff and Patriarch Evans that while most victims will contract the plague by contact with the compound through the skin or lungs (producing hemorrhaging and eventual death), those who ingest the contaminated water will suffer more horribly still. When taken into the mouth, the compound will cause the cells of the body's soft tissues in and around the mouth to literally rupture, as Patriarch Evans described above, causing the tongue to appear to "dissolve" and the eyes to "rot" and "fall" from their sockets. Some of the flesh would literally "fall from off their bones."

An analysis of the effects of the lethal compound

presents a consistent diagnosis. It causes the cells of the body to rupture; light concentrations on the skin induce massive hemorrhaging, while heavier concentrations cause a total dissolution of tissue. Either way, a painful death follows.

The message comes through loud and clear: make no contact with the compound that will precipitate from the sky, nor with the water it will contaminate!

Saints generally spared

As a further matter of interest to Latter-day Saints, in the vision recorded by Elder Woodruff, the inhabitants of Salt Lake also suffered from the effects of the plague. But instead of seeing the dead and dying everywhere, as in other parts of the country, he saw a somewhat more hopeful picture:

> I was immediately in Salt Lake City wandering about the streets in all parts of the city, and on the door of every house I found a badge of mourning, and I could not find a house but what was in mourning . . . . It seemed strange to me that I saw no person on the street in my wandering about through the city. They seemed to be in their houses with their sick and dead. 'I saw no funeral procession, or anything of that kind, but the city looked very still and quiet as though the people were praying, and had control of the disease whatever it was. (Op. cit.)

It appears that the residents of Salt Lake will know how to control the plague, although there will be death in every family; and they will have taken the most prudent action to avoid contamination by staying indoors.

With this understanding of the nature of the plague, it comes as no surprise that the streets of Salt Lake were empty. The best way to avoid contact with the extraterrestrial compound which would fill the air would be to remain indoors. Remember the advice Isaiah gave to those who would wish to survive the "indignation" of the Lord:

> Come, my people, enter thou into thy chambers, and shut thy doors about thee: hide thyself as it were for a little moment, until the indignation be overpast. (Isaiah 26:20.)

Logic dictates that the safest place during the lethal dustfall is indoors. Isaiah was giving sound advice.

The question immediately comes to mind: why would the people in Salt Lake know how to deal with the plague better than the people in New York, Philadelphia, or Boston who were also suffering and dying? Answer: just as the Israelites who obeyed Moses were prepared and able to cope with the plague, so will the Saints (and those who live among them) survive under the leadership of an inspired prophet who will warn them and give precautions.

However, even with such precautions, it appears that the illness will take many. Elder Woodruff recorded there were many fatalities among the Saints: "on the door of every house I found a badge of mourning. . . ."

Can we prepare?

While there may be little we can do to avoid the other plagues of the last days because of their devastating nature, this plague—deadly though it may be—is survivable. Wisdom would dictate that stored food and water are our most valuable assets in a plague such as that described in this chapter. We would be isolated from even our closest neighbors by the invading dust. Because of the need to remain indoors and away from the deadly dust, stored food and pure water would be mandatory for survival. One could not sally forth to secure either—even if it were available—without risk of sudden death. Food and water stored in the home would be excellent insurance against this plague, and it is most fascinating to discover that the things we have been counseled to store by the Brethren would be the very things needed to survive such a plague. We can begin our preparations by following this counsel.

"For the hour is nigh and the day soon at hand when the earth is ripe; and all the proud and they that do wickedly shall be as stubble; and I will burn them up, saith the Lord of Hosts . . . ."

Doctrine & Covenants 29:9

13

Fire From Heaven

The Lord through His prophets has had a few things to say about the fire and hail of hot ash and stones that will fall from heaven in the last days. The Prophet Joel spoke of the invading army of God, the hosts of heaven, and said that their arrival would be accompanied by a terrible, literal burning:

> A fire devoureth before them; and behind them a flame burneth: land is as the garden of Eden before them, and behind them a desolate wilderness; yea, and nothing shall escape them. (Joel 2:3.)

In Revelation, John described the same army and called them "locusts." He also maintained that the principle weapon of that army was fire, accompanied by smoke and brimstone:

> . . . and out of their mouths issued fire and smoke and brimstone.

> By these three was the third part of men killed, by the fire, and by the smoke, and by the brimstone, which issued out of their mouths. (Revelation 17:18.)

From space

Of course, Velikovsky attributed the fire that fell from heaven anciently to hydrocarbon-based materials that entered Earth's atmosphere during planetary close encounters or when Earth encountered the orbiting debris created in those catastrophes. Evidently the same cause and effect relationship will exist in future encounters prophesied to occur in the last days.

John wrote that the number of the army he saw in vision was "two hundred thousand thousand." A large comet-like body torn asunder by a great near-collision would fill space with incredible amounts of solid and gaseous material. The gases would be of many kinds—some inert, others quite flammable. The solid material would range from the finest grains to chunks the size of giant boulders, to moon-sized pieces. Such a collision would create millions of smaller comet-like bodies. Those bodies, together with other debris and gases, would move around the Sun, strung-out along an orbit similar to that of their parent body. Each time Earth's orbit converged with the orbit of the debris, Earth would plunge through it, causing prodigious amounts of material—gaseous and solid—to pour into our atmosphere. Such an event would be most spectacular and destructive, as well as impressive.

John, in the ninth chapter of Revelation, and Joel in the Old Testament both seem to be describing (metaphorically) just such an occurrence among the catastrophes of the last days. When read from this perspective, the books of Joel and Revelation become much more plain and understandable.

The experience gained through the use of petroleum products in modern times allows a better understanding of the damage that can be done by flammable hydrocarbons. Natural gas, seeping into the air, mixing with the oxygen, and then igniting, has done tremendous damage. Refinery and oil-well fires have burned out of control for days and sometimes weeks at a time while men have been unable to do anything but watch them burn.

If, as Velikovsky suggested, strings of burning fluid poured onto the surface of the Earth (as in past close planetary encounters), then we may literally see pools of burning fluid pouring onto the surface of the Earth in scattered locations. And if the descent of this flammable liquid were to be accompanied by falling meteorites (as it was in Egypt during the Exodus), igniting the liquid, then the Prophets' metaphor of "a lake of fire and brimstone" may become an awesome reality.

Lest any say it is unlikely such flammable material could come from the sky, remember that hydrocarbons are present in immense quantities in the atmospheres of many of the planets and moons in the solar system. The recent discovery of the high concentrations of hydrocarbons in other atmospheres has been rather astonishing to scientists. Someone has said that it is remarkable that our solar system has not yet exploded because of its abundance of flammable elements. Certainly, any mixing of Earth's atmosphere (having a high concentration of oxygen necessary for combustion) with that of another planet or moon (containing flammable hydrocarbon compounds) would have disastrous results. But the fire that will fall from heaven in the last days will not be just burning gases or fluids but also hot ash and stones: a formidable way of destroying the flora and fauna of Earth.

Prophesied burnings

Latter-day prophets have seen this fire from heaven in their visions of the last days. Patriarch Charles Evans saw and described a fire that will accompany the other effects of a future catastrophe:

> In other parts voluminous flames, emanating from vast fires, rolled with fearful velocity destroying life and property in their destructive course. (*The Contributor,* Vol. 15, No. 20, p. 638.)

In the vision recorded by Elder Woodruff, the City of New York was destroyed by fire after its population was decimated by the great, red plague from heaven:

> I then went to the Grand Central Park [New York] and, looking back, I saw a fire start and just at that moment a mighty east wind sprang up and carried the flames west over the city, and it burned until there was not a single building left standing whole, even down to the wharfs. And the shipping all seemed to be burned, and swallowed up in the common destruction and left nothing but a desolation where the great city was a short time before. (*Wilford Woodruff Journals,* June 15, 1878.)

Notice that this is no ordinary fire. It more nearly resembles the firestorms created by the Allied incendiary bombs in World War II, when they were dropped en masse from airplanes on the cities of Dresden and Hamburg. But there is no mention of hearing or seeing any explosions, so that would seem to rule out bombs (nuclear or otherwise). A nuclear blast would vaporize the city almost instantly—a much different effect than the fire that was seen. (Notice that the ships were burned down to the waterline, not blasted out of the water.)

What could cause such an intense fire to spring up—almost spontaneously? In all likelihood, what is being described is the effect of flammable materials which will be introduced into Earth's atmosphere by a passing body. This "baptism by fire" has been described by virtually every prophet that has seen the last days in vision. Both the visions cited describe these fires in conjunction with the other signs of global catastrophe, a further indication that what was seen in vision was the prophesied "burnings" of the last days.

A small preview

Interestingly, there is an example in rather recent times (historically speaking), of a fire which seemed to have a life of its own, springing up almost spontaneously until nearly the entire city was consumed—the great Chicago fire of 1871. What is more, there are strong indications that that fire holds a clue to the nature of the coming holocaust.

Though folk tradition points to Mrs. O'Leary's cow as the culprit responsible for the fire that literally destroyed Chicago, more thoughtful inquirists were unable to find the exact cause. "Spontaneous combustion" was the cause most often cited for lack of any other cause. None of those who investigated the fire actually believed that O'Leary's cow started the fire!

The fire-marshall testified:

> I felt it in my bones that we were going to have a burn . . . .
>
> We got the fire under control, and it would not have gone a foot farther; but the next thing I knew they came and told me that St. Paul's church, *about two squares north, was on fire . . . .*
>
> The next thing I knew the fire was in Bate-

ham's planing-mill." (*History of the Great Conflagration,* p. 163.)

Eyewitness accounts of the fire tell of its strange nature:

> The huge stone and brick structures melted before the fierceness of the flames as a snow-flake melts and disappears in water, and almost as quickly. Six-story buildings would take fire and *disappear for ever from sight in five minutes by the watch* . . . . The fire also doubled on its track at the great Union Depot and burned half a mile southward *in the very teeth of the gale*—a gale which blew a perfect tornado, and in which no vessel could have lived on the lake. . . . *Strange, fantastic fires of blue, red, and green played along the cornices of buildings.* (*History of the Chicago Fire,* pp. 85, 86.)

> The fire was accompanied by the fiercest tornado of wind ever known to blow here. (Ibid., p. 87.)

> The most striking peculiarity of the fire was its intense heat. Nothing exposed to it escaped. Amid the hundreds of acres left bare there is not to be found a piece of wood of any description, and, *unlike most fires, it left nothing half burned. . . .*

> The fire swept the streets of all the ordinary dust and rubbish, consuming it instantly. (Ibid., p. 119.)

No ordinary fire sweeps the debris from the streets. No ordinary fire consumes everything, "leaving nothing half burned." Even the firestorms in German cities, created by incendiary bombs (and the atomic blasts in Japan), left charred remains of the devastated cities.

No ordinary fire advances into the wind, indeed, "in the very teeth of the gale." The fires in Chicago were not being advanced by the wind—they moved into the wind. The explanation for such a phenomenon is that the incendiary agent that was igniting the fires was being carried by the wind itself.

But that is not all! The story becomes more amazing still. What history neglects to tell us is that the great fire in Chicago was but a small part of a great conflagration that spread over almost all of North America. Ignatius Donnelly, an author who published after the fire (1883) and who lived in the vicinity at the time, wrote:

> In the year 1871, on Sunday, the 8th of October, at half past nine o'clock in the evening, events occurred which attracted the attention of the whole world, which caused the death of hundreds of human beings, and the destruction of millions of [dollars worth of] property, and which involved three different States of the Union in the wildest alarm and terror . . . .
>
> At that hour, half past nine o'clock in the evening, *at apparently the same moment,* at points hundreds of miles apart, in three different states, Wisconsin, Michigan, and Illinois, fires of the most peculiar and devastating kind broke out, so far as we know, by spontaneous combustion. (*Ragnarok.*)
>
> It must not be forgotten that the fall of 1871 was marked by extraordinary conflagrations in regions widely separated. On the 8th of October, *the same day* the Wisconsin, Michigan, and Chicago fires broke out, the states of Iowa, Minnesota, Indiana, and Illinois were severely devastated by prairie-fires; while terrible fires raged on the Alleghenies, the Sierras of the Pacific coast, and the

> Rocky Mountains, and in the region of the Red River of the North. (Ibid.)
>
> For weeks before and after the great fire in Chicago in 1871, great areas of forest and prairie-land, both in the United States and the British Provinces [Canada], were on fire. (*The Annual Record of Science and Industry*, 1876, p. 84, as quoted in *Ragnarok*.)

Unlike other investigators, Donnelly did not believe that the fires began spontaneously. Why would a major portion of the continent suddenly begin to burn? Two or even three fires could conceivably begin spontaneously and simultaneously in widely separated areas, but not as many widespread fires as occurred in 1871-72.

Extraterrestrial source

Donnelly attributed the holocaust to flammable gas and debris from the tail of Biela's comet through which Earth had just passed. Biela's comet was a disturbed and spectacular body which first appeared in 1826. It's orbit was irregular and menaced Earth every 6-7 years. *On at least one occasion it crossed Earth's orbit only one month ahead of the Earth.* In 1846 something caused the comet to split in two and dismembered its tail.

Biela's comet was far too small to produce the many effects of ancient cataclysms, but it did have a large tail which, likely, contained gases and debris similar to the bodies that did such damage anciently. In fact, Velikovsky theorized that many of the comets in existence today are remnants of past planetary catastrophes. Apparently in 1871 Earth penetrated the disconnected tail of Biela's comet: flammable gases and fine debris penetrated Earth's atmosphere and descended over North America. The destruction which followed may well be a mild foretaste of what is to come in the "burnings" of the last days:

In Wisconsin, on its eastern borders, in a heavily timbered country, near Lake Michigan, a region embracing *four hundred square miles,* extending north from Brown County, and containing Peshtigo, Manistee, Holland, and numerous villages on the shores of Green Bay, was swept bare by an absolute whirlwind of flame. There were *seven hundred and fifty people killed outright,* besides great numbers of the wounded, maimed, and burned, who died afterward. More than three million dollars' worth of property was destroyed. (*History of the Great Conflagration,* pp. 393, 394.)

At sundown there was a lull in the wind and comparative stillness. For two hours there were no signs of danger; but at a few minutes after nine o'clock, and by a singular coincidence, *precisely the time at which the Chicago fire commenced,* the people of the village heard a terrible roar. It was that of a tornado, crushing through the forests. *Instantly the heavens were illuminated with a terrible glare. The sky,* which had been so dark a moment before, *burst into clouds of flame.* A spectator of the terrible scene says the fire did not come upon them gradually from burning trees and other objects to the windward, but the first notice they had of it was *a whirlwind of flame in great clouds from above the tops of the trees,* which fell upon and entirely enveloped everything. The poor people inhaled it, or the intensely hot air, and fell down dead. . . .

This phenomenon seems to explain the fact that so many were killed in compact masses. They seemed to have huddled together, in what were evidently regarded at the moment as the safest places, *far away from buildings, trees, or other inflammable*

> material, and there to have died together." (Ibid., p. 372.)
>
> It is the universal testimony that the prevailing idea among the people was, that the last day had come. Accustomed as they were to fire, nothing like this had ever been known. They could give no other interpretation to this ominous roar, this *bursting of the sky with flame, and this dropping down of fire out of the very heavens,* consuming instantly everything it touched.
>
> No two give a like description of the great tornado as it smote and devoured the village. It seemed as if 'the fiery fiends of hell had been loosened,' says one. 'It came in great sheeted *flames from heaven,*' says another. 'There was *a pitiless rain of fire and SAND.*' 'The atmosphere was all afire.' Some speak of *'great balls of fire unrolling and shooting forth in streams.'* The fire leaped over roofs and trees, and ignited whole streets at once. No one could stand before the blast. It was a race with death, above, behind, and before them. (Ibid., p. 374.)

The above descriptions are at once reminiscent of the Exodus account, mythological accounts of fire from heaven, and prophetic visions of the last days. If Donnelly was correct in connecting these unusual fires to the flammable gases from the tail of Biela's comet, then the above accounts are a small, *but accurate,* example of what will happen in the future when Earth encounters much larger bodies in space, capable of far greater devastation. What is more, the prophetic accounts describing the latter-day conflagration become more realistic and ominous.

(For a more complete account of Donnelly's assessment of the fires from heaven, see Appendix A.)

". . . and the earth trembled, and the mountains fled, even according to his command; and the rivers of water were turned out of their course . . . and all nations feared greatly . . . . And from that time forth there were wars and bloodshed among them . . ."

Moses 7:13, 16

14

Wars and Rumors of Wars

One of the indirect effects of the coming catastrophes will be the change in behavior it will bring to mankind. An awareness of those changes will undoubtedly aid us in our struggle should we survive.

Because of the deprivation and hunger that will result from the coming catastrophes, some survivors will feel compelled to leave the area around them which will have been devastated. Thinking that their land alone suffered the wrath of God and not realizing that other lands have suffered the same catastrophes, these bands of refugees will migrate in search of undamaged crops, potable water, etc. The compulsion to *migrate* in search of food and water will naturally bring them in conflict with other groups, and warfare will result. Such was the case anciently during cataclysms.

Exodus war

After the Israelites had left Egypt, they encountered the Amalekites.

> Then came Amalek, and fought with Israel in Rephidim . . . .
>
> And Moses said unto Joshua, Choose us out men, and go out, fight with Amalek; tomorrow I will stand on the top of the hill with the rod of God in mine hand . . . .
>
> And it came to pass, when Moses held up his hand, that Israel prevailed; and when he let down his hand, Amalek prevailed.
>
> . . . and Aaron and Hur stayed up his hands, . . . and Joshua discomfited Amalek and his people with the edge of the sword. (Exodus 17:8-9, 11, 12, 13.)

According to Velikovsky, Arabic sources describe how the Amalekites had "left Hedjaz because of plagues, [and] followed the cloud in their wandering through the desert." (*Worlds In Collision,* p. 142.) He asserted that they, too, had been driven from their lands by the catastrophe. "The fear of the Judgment Day not only did not pacify the nations, but on the contrary, uprooted them, impelling them to migration and war." (Ibid., p. 270.) Hence, in the encounter between the Israelites and the Amalekites, we see a picture of two wandering nations, driven from the devastation in their homeland, encountering one another, and doing battle in the desert. (Just as a note of interest, Velikovsky indicated that the Amalekites were a powerful nation. After doing battle with the Israelites, they continued on into Egypt, subjugated the hapless Egyptians, and ruled for many years. They are known to profane history as the Hyksos!)

Madness

But it would appear that such belligerence can be attributed to more than simply a struggle for survival. Dr. Velikovsky, referring to ancient accounts of violence, noted that: "Civil war in the nations, tribal strife, and strife between members of households became so widespread that the same complaint was heard in many parts of the world." (Ibid., p. 270.)

He cited Egyptian records:

> While years passed in hostility, each one seizing upon his neighbor, not remembering his son to protect. (Breasted, *Records of Egypt,* as quoted in *Worlds In Collision,* p. 270.)
>
> I show thee the land upside down; the sun is veiled and shines not in the sight of men. I show thee the son as enemy, the brother as foe, a man slaying his father. (Gardiner, "New Literary Works from Ancient Egypt," *Journal of Egyptian Archaeology,* as quoted in *Worlds In Collision,* p. 271.)

The described conflicts were not between nations, but between individuals—even between members of the same family. But there seems to be more at work in these conflicts than the instinct for survival. What madness causes murder and mayhem between neighbors and even family members?

After analyzing descriptions in ancient accounts of abnormal behavior, Dr. John V. Meyers (Language Department Chairman, Union University) and Warner B. Sizemore (Assistant Professor of Religion, Glassboro State College) wrote: "the shattering and prolonged trauma of cosmic catastrophe could produce psychological states which were strikingly similar to those which we classify today as catatonic schizophrenia." (*Kronos,* Vol. 1, No. 1, p. 66.)

Apparently the overwhelming nature of those events left the participants no alternative but to withdraw from the awful reality of what was happening to them. Unable to cope with the devastating nature of the cataclysm, the human mind turned to a mental defense mechanism—catatonic schizophrenia—in an attempt to shut out the incomprehensible changes occurring in the physical world. A Ph.D., and counselor, Thomas Allen Parry noted:

> On every side man would be inundated with a rain of destruction and danger far too great for his normal capacities to cope with. The active mode of response would be immobilized, for there would be virtually nothing that could be done to control or even escape the situation. Only the receptive mode could operate, for the organism, both individual and collective, could only receive what was happening. In order to discern the meaning of it all, or even just to stay alive, formal language, logical sequences of events, assumptions of constancy and rationality would be hopelessly inadequate. . . .
>
> It may be supposed that the events themselves were so overwhelming that there was no room for exaggeration, only an awestruck taking in of the spectacle which was inundating people. (*Kronos,* Vol. I, No. 1, pp. 7, 14.)

The textbook definition of catatonic schizophrenia describes a loss of contact with reality, which results in two types of behavior. The first is a mute type of the affliction: the individual withdraws into a state of absolute silence and isolation. "The patient may maintain difficult postures for hours, days, or months and requires complete care." (*Abnormal Psychology,* p. 316.) The second is catatonic excitement or furor in which the individual becomes maniacal. He runs amok—he literally goes "crazy." "In the excited state, he becomes extremely agitated and destructive. He is likely to destroy furniture, tear his clothes, assault others, or injure and mutilate himself." (Ibid., p. 316.) In extreme cases of this hyperactive, agitated state, the individual will attempt to kill or maim anyone he can reach. Since family or friends, neighbors, and associates are often the most accessible to the "mad-man," they become his most likely victims. Hence the ancient accounts tell of conflicts between those most closely associated.

This conclusion carries tremendous implications for those who wish to survive the coming catastrophes. If mankind is to suffer from catatonic schizophrenia during coming cataclysms, then we would be well advised to learn all we can of its symptoms and how to deal with them in ourselves and in others.

Prophecy is replete with warnings of this type of behavior, similar to that exhibited anciently during catastrophic episodes. Most seem to prophesy of a state of catatonic furor:

> And it shall come to pass in that day, that a great tumult from the Lord shall be among them; and they shall lay hold every one on the hand of his neighbour, and his hand shall rise up against the hand of his neighbour. (Zechariah 14:13.)
>
> And I will overthrow the throne of kingdoms, and I will destroy the strength of the kingdoms of the heathen; and I will overthrow the chariots, and those that ride in them; and the horses and their riders shall come down, every one by the sword of his brother. (Haggai 2:22.)
>
> And there shall be earthquakes also in divers places, and many desolations; yet men will harden their hearts against me, and they will take up the

> sword, one against another, and they will kill one another. (Doctrine & Covenants 45:33.)
>
> I have sworn in my wrath, and decreed wars upon the face of the earth, and the wicked shall slay the wicked, and fear shall come upon every man; (Doctrine & Covenants 63:33.)
>
> . . . no man shall spare his brother. (Isaiah 9:19.)
>
> . . . every man's sword shall be against his brother. (Ezekiel 38:21.)

The clearest statement to the psychological effect of great catastrophes comes from the Prophet Joseph Smith:

> I saw men hunting the lives of their own sons, and brother murdering brother, women killing their own daughters, and daughters seeking the lives of their mothers. (*History of the Church,* Vol. 3, p. 390.)

The same kind of madness among the people was seen in the vision recorded by Elder Woodruff:

> The states of Missouri and Illinois were in turmoil and strife, men killing each other, and women joining in the fight, family against family, cutting each other to pieces in the most horrid manner.
>
> . . . and they were crazy mad, and in a dying condition (*Wilford Woodruff Journals,* June 15, 1878.)

Notably the words "crazy mad" were used to describe the behavior of those seen in the vision of the future. (Please note that this madness is not an isolated phenomenon. It occurs simultaneously with the onset of the natural destructions and the celestial apparitions.)

Zephaniah wrote a rather accurate, if incomplete,

description of the other manifestation of this mental disturbance—the slow and shuffling locomotion of the mute, but mobile, catatonic schizophrenic:

> The great day of the Lord is near . . . .
>
> That day is a day of wrath, a day of trouble and distress, a day of wasteness and desolation, a day of darkness and gloominess, a day of clouds and thick darkness . . . .
>
> And I will bring distress upon men, that they shall walk like blind men . . . (Zephaniah 14-15, 17.)

Descriptions in prophecy of both types of catatonic schizophrenia among men leaves little doubt about the profound mental afflictions which will accompany the physical dangers that will plague mankind during future encounters.

Apparently both men and animals will be afflicted:

> In that day, saith the Lord, I will smite every horse with astonishment, and his rider with madness: and I will open mine eyes upon the house of Judah, and will smite every horse of the people with blindness. (Zechariah 12:4.)

Because of the declaration of "madness" among men at this period, it does not seem unlikely that the reference to blindness refers to that psychosomatic affliction called hysterical blindness—a likely side effect of an overwhelming catastrophe.

Though the picture of latter-day destructions, plagues, and madness is not a pleasant one to contemplate, such knowledge would undoubtedly be of immense value to those who are called upon to bear that burden. If they can recognize the changes of behavior occuring, they will be better prepared to deal with them.

Should it be our lot to endure the hardships of those days, understanding the nature of the overwhelming manifestations and knowing how to cope with them emotionally may mean the difference between life and death.

"Then shall be fulfilled that which is written, that in the last days, two shall be in the field, the one shall be taken, and the other left; Two shall be grinding at the mill, the one shall be taken, and the other left . . ."
Joseph Smith—Matthew 1:44, 45

15

Prophets of Doom—Prophets of Hope

After reading of the horrible, world-wide, earth-shaking destructions enumerated in the prophetic books, the reader may become overwhelmed by the depressing aspects of such events. Such an impression comes easily when dwelling on the destructions of the last days. With the insight into history and prophecy provided by Dr. Velikovsky's theories, the coming events take on an even darker aspect, a more foreboding ambiance than ever before.

The Prophet Joseph Smith also appears to have commented on this aspect of prophecy. He has been quoted as saying:

> Let us reflect, then, in the last days, that there was to be great tribulations; for the Savior says, nation shall rise against nation, kingdom against kingdom, and there shall be famine, pestilence, and earthquakes in divers places, and the prophets have declared that the valley should rise; that the mountains should be laid low; that a great earthquake should be, in which the sun should become black as sackcloth of hair, and the moon turn into blood; yea, the Eternal God hath declared that the great deep shall roll back into the north countries and that the land of Zion and the land of Jerusalem shall be joined together, as they were before they were divided in the days of Peleg. *No wonder the mind starts at the sound of the last days.* (*Inspired Prophetic Warnings,* pp. 41, 42, italics added.)

Why such calamities?

Orson F. Whitney may have echoed our thoughts when he wrote, "And what is the purpose—the ultimate purpose of it all? Destruction?" With an enhanced understanding of prophecy and the events to befall mankind in these last days, the question of "Why?" is one we might all ask ourselves. Why, indeed? Is man so insignificant and his sufferings so miniscule that they are of no consequence to the Creator? Is mass destruction truly the way of a loving Father? Elder Whitney answered:

> No, a thousand times no, except insofar as destruction must at times precede reconstruction, . . . The world's welfare is the object in view. . . . However severe his chastisements, we can rest assured of this: Hatred of humanity has no place in the heart of him who "so loved the world" that he "gave his only begotten Son" to save it from eternal

> damnation.
>
> Calamities do not come upon the world merely to scourge the wicked and avenge the wrongs of the righteous. The primal aim of divine punishment is to purify . . . . Wars and other woes are sent to put a stop to men's evil practices, lest they add sin to sin and pile up guilt to their greater condemnation. To be swept off the earth and ministered to in the spirit world is not the worst fate that can befall the wicked. *Omnipotence wields the powers of destruction in such a way as to make of them instruments of salvation.* It may seem cruel, but in reality, it is kind. (*Saturday Night Thoughts,* p. 203, italics added.)

Though trying to impress upon us the profoundly devastating nature of future events, it is likely that the purpose of the prophets in doing so was *to prepare us,* not frighten us. Though they certainly hoped that a healthy respect for the power and might of the Lord might persuade us to repent of our misdeeds, it is doubtful that the prophets elaborated on all the dire consequences of celestial catastrophes in order to sensationalize them. In fact, the prophets must have been overwhelmed by the enormity of the events they experienced personally and in vision. Logically, then writing or speaking about such profoundly devastating events, one would think that they might be inclined to elaborate and explore each facet or detail. Considering the nature and extent of global catastrophies, it is surprising that the prophets managed to relate those events in such a matter-of-fact manner. However, it must be remembered that the calling of the prophets is to warn the world and call them to repentance, not to dwell upon the woeful aspects of such catastrophes.

One of the cardinal rules of life is that the old order of things must be removed in order to initiate the new. The

Earth must be "renewed" in order to receive its "paradisaical glory." In today's "urban renewal" projects, the old buildings are demolished to allow new structures to be built in their place. In a like manner, the Lord will destroy the old Earth in order to make way for the new Earth. Peter used the words "new heaven" and "new earth" to describe the changes to take place. Thus, the world will be vastly changed to prepare it for its millennial rest.

No surprises

Despite the ominous tone of the prophesies of destruction, it appears that the Lord's intent is to inform and prepare His Saints. Of those Thessalonians who had embraced the gospel, repented, and prepared themselves for what was to come by learning the signs, Paul wrote the following:

> But of the times and the seasons, brethren, ye have no need that I write unto you.
>
> For yourselves know perfectly that the day of the Lord so cometh as a thief in the night.
>
> For when they shall say, Peace and safety; then sudden destruction cometh upon them, as travail upon a woman with child; and they shall not escape.
>
> But ye, brethren, are not in darkness, that that day should overtake you as a thief.
>
> Ye are all the children of light, and the children of the day: we are not of the night, nor of darkness.
>
> Therefore let us not sleep, as do others; but let us watch and be sober [vigilant]. (1 Thessalonians 5:1-6.)

Paul clearly felt that the Saints in his day should not be taken by surprise by the events of the last days. "But ye, brethren, are not in darkness, that that day should overtake you as a thief." He implied that the Saints should be well enough schooled in the gospel that their house would be set in order and that they should know exactly what to expect in "the day of the Lord." Obviously, the worldly will be taken by surprise, but the Saints should not be.

Prepare for the last days

At first glance, it may seem that there is little one could do to prepare for the coming destructions. The puny efforts of man cannot withstand the onslaught of nature's elements gone wild. However, there were always those who survived the catastrophes of the past. As Velikovsky observed, "We are the descendants of survivors, themselves descendants of survivors." The Lord has promised that He will deliver the more righteous part from the impending destructions. In addition, while painting a gloomy picture for the immediate prospects of mankind, prophecy also predicts a future of beauty and serenity in a millennial world, renewed and glorified, where men and women will live to a ripe old age—a world well worth living to see. So, while we may be certain that there will be terrible destructions that will destroy the majority of humanity, there will also be many who will survive to see a wonderful new age ushered in—*an experience worth living for.*

A hope for survival

When the prophets told how we might prepare for the last days, they spoke in terms of personal righteousness and belief in salvation through the Messiah. The message of the scriptures (nothwithstanding the calamities they enumerate) is a positive message—a message of hope. Prophecy is given so that we might know what to expect, and we have the gospel, the scriptures, and living prophets to direct us so that we might know how to prepare for what is to come. This preparation and the opportunity for salvation and survival is available to all men and women.

There is an important caution to observe in pursuit of survival. After all the precautions and all the care we might take, nothing we can do will free us completely from the possibility of death in the coming catastrophes. The destruction of past cataclysms has been most general and somewhat indiscriminate in nature; future destructions, in all likelihood, will be equally general and indiscriminate. Perhaps that is why Joseph Smith wrote:

> . . . it is a false idea that the Saints will escape all the judgements, whilst the wicked suffer; for all flesh is subject to suffer, and "the righteous shall hardly escape;" still many of the Saints will escape, for the just shall live by faith; yet many of the righteous shall fall a prey to disease, to pestilence, etc., by reason of the weakness of the flesh, and yet be saved in the Kingdom of God. (*History of the Church*, Vol. 4, p. 11.)

So, it appears that we cannot absolutely insure our survival in the coming catastrophes—even righteousness is no guarantee of survival. A righteous life guarantees salvation and even exaltation; it does not guarantee survival in the flesh. However, upon closer examination it becomes evident that along with obedience to the laws of God there are several things that may considerably improve our chances of survival.

Follow the prophet

Anciently, during events very similar to those we face in the near future, the direction of an oracle of God was invaluable. When God informed Moses, for example, of the destructions to be poured out upon Egypt, his advice to the

Israelites was specific, telling them what they might do to preserve themselves from the impending doom. Moses instructed them to stay indoors during the darkness and earthquake of the passover. (Exodus 12:21-22.) Also, logic tells us that Moses undoubtedly informed his fellow Israelites of each impending plague just as he informed Pharaoh.

Another example from Exodus of the importance of following the prophet comes from the event known as the parting of the Red Sea. After leaving Egypt, upon receiving instructions from Moses, the Israelites crossed the Red Sea—something they would not have known to do or would have feared doing on their own, without divine direction. (Exodus 14:15-16, 21-22.) As it turned out, the Egyptians' attempt to imitate the Israelites by also crossing the Red Sea was fatal, as it would have been for anyone lacking divine guidance. In this instance, timing was critical—the sea had only been parted temporarily. Moses, in and of himself, would have had no way of knowing beforehand what to expect or what to do unless he were instructed by God. The most logical emotional response to such a phenomenon on the part of Moses and the Israelites would have been fear and flight! To the best of my knowledge, no one before that time had ever done such a thing—there was no precedent for Moses and the Israelites to follow. However, the Lord knew how to take advantage of the moment, and directed Moses to lead the Israelites across the open seabed to safety. And so it was that Jehovah directed the House of Israel, through Moses, that they might be preserved from death, while others all around them, who were not obedient to the words of the prophet, suffered and died.

The lesson to be learned, then, is simply this: follow the prophet! This is certainly not new advice to Latter-day Saints, but it does serve to explain how those who had sufficient faith to follow the prophet and the Priesthood were spared in past catastrophes and how we today may also be spared the worst that the future has to offer. Undoubtedly, this is the means whereby the Lord plans to fulfill His promise to spare the righteous: the righteous will follow the prophet who is guided by the wisdom and foreknowledge of God, and they will thereby avoid the worst of the impending plagues and destructions.

Not enough?

Some may say that it would suffice us to simply leave it at that. They would say that it is enough to know that we can depend upon the words of the prophet and live. However true that may be, our faith in the prophet's leadership is enhanced when we understand, at least to some degree, *why* he does what he does. For that reason we are given the scriptures that we may have a testament of truth from the past to aid us to learn to exercise our faith properly. Given all the information in prophecy, it certainly appears that the Lord intended us to understand the nature and sequence of coming events. If we understand the prophetic outline of the future, then we will better understand what motivates the prophet as he guides the righteous. Understanding why the prophet takes a certain action gives us greater motivation to follow him, and allows us to better serve others who may need our support.

In addition, a foreknowledge of the nature of upcoming cataclysms will be invaluable to those who wish to survive to see the Millennium ushered in. The human mind, heart, and spirit can withstand tremendous hardships if properly prepared and conditioned in advance. For that very reason soldiers in training are subjected to tests and trials in order to prepare them for the adversity of battle. Such conditioning does not make the coming hardships less difficult, but it does improve our ability to deal with them, making them less difficult to bear. It simply prepares the individual as to what to expect so they are not overwhelmed in the midst of the struggle. And make no mistake, men will be tested to the extreme in the coming catastrophes. An eminent scholar of the scriptures and of prophecy, Cleon Skousen wrote:

> In contemplation of our dispensation the prophets reached across the centuries to tell us of the great events that would mark our day among the seasons of eternity . . . . In the program of the Lord, we have a part—it may be a most important part. The prophets wrote, therefore, to warn us and to teach us that we might perform our parts intelligently and well. (*Prophecy and Modern Times,* pp. 11-12.)

"To be forewarned is to be forearmed." That seems to be the underlying purpose in all prophecy. Hence, it is prudent to examine past events and prophesies of the future in order to learn all we can of the obstacles ahead.

. . . *And The Earth Shall Reel To And Fro. . .*

Appendix A

A most amazing discovery awaited me when I began looking into events that have occurred since the restoration of the Gospel which might be construed as signs of the last days. The discovery was this: in recent times mankind has been given a foretaste of calamities to come. Even in modern times, our course through the solar system has not been without incident.

The following is an excerpt from a fascinating book by an early catastrophist, Ignatius Donnelly, entitled *Ragnarok: The Age of Fire and Gravel.* (Published in 1883 in New York by D. Appleton and Company, it is no longer in print.)

Donnelly preceded Velikovsky by nearly seventy years, yet he had concluded that the Earth suffered tremendous cataclysms in early historic times due to passing comets. In that respect he was an acknowledged precursor to Velikovsky. But because Donnelly was not as thorough, nor as well trained as Velikovsky (nor did he have access to the wealth of knowledge Velikovsky did), he misread much of the evidence and reached some erroneous conclusions.

Nevertheless, he made many interesting points. The most startling evidence that he presented concerned a phenomenon that occurred in his own lifetime. Rather than paraphrase and comment on the entire contents of chapter 5 from his book, it seemed more useful to allow the reader to study Donnelly's own words. It is this author's belief that what he had to say will be fascinating to the reader.

- - - - - - - - - - -

There is reason to believe that the present generation [1883] has passed through the gaseous prolongation of a comet's tail, and that hundreds of human beings lost their lives, somewhat as they perished in the Age of Fire and Gravel [Donnelly's name for the time when Earth and comets nearly collided], burned up and poisoned by its exhalations.

And, although this catastrophe was upon an infinitely smaller scale than that of the old time, still it may throw some light upon the great cataclysm. At least it is a curious story, with some marvelous features:

On the 27th day of February, 1826, (to begin as M. Dumas would commence one of his novels), M. Biela, an Austrian officer, residing at Josephstadt, in Bohemia, discovered a comet in the constellation Aries, which, at that time, was seen as a small round speck of filmy cloud. Its course was watched during the following month by M. Gambart at Marseilles and by M. Clausen at Altona, and those observers assigned to it an elliptical orbit, with a period of *six years and three quarters* for its revolution.

M. Damoiseau subsequently calculated its path, and announced that on its next return the comet would cross the orbit of the earth, within *twenty thousand miles of its track,* and but about *one month before the earth would have arrived at the same spot!*

This was shooting close to the bull's-eye!

He estimated that it would lose nearly ten days on its return trip, through the retarding influence of Jupiter and Saturn; but, if it lost forty days instead of ten, what then?

But the comet came up to time in 1832, and the earth *missed it by one month.*

And it returned in like fashion in 1839 and 1846. But here a surprising thing occurred. *Its proximity to the earth had split it in two;* each half had a head and tail of its own; each had set up a separate government for itself; and they were whirling through space, side by side, like a couple of race-horses, about sixteen thousand miles apart, or about twice as wide apart as the diameter of the earth . . . .

In 1852, 1859, and 1866, the comet SHOULD have returned, but it did not. It was lost. It was dissipated. Its material was hanging around the earth in fragments somewhere. [More likely, it was moving in nearly the same orbital path; but, as Donnelly later makes the point, its tail had been stolen away and dismembered—hence it went undetected by astronomers until 1872.]

[At this point in the narrative Donnelly cites the evidence for the fact that the intersection of Earth and Biela's comet in 1872 was announced by a spectacular meteoric display in Earth's atmosphere. Astronomers had been vainly searching for Biela, and the meteor shower caused them to suspect that it was somewhere in the vicinity of the Earth. Indeed, with the aid of this clue, the comet was soon sighted, moving away from the Earth. The meteor shower had been the result of Earth narrowly missing Biela and colliding with whatever remains there were of its tail traveling in the track of the comet. There is no question about this order of events; it is a matter of historical verity.]

But did the earth escape with a mere shower of fireworks?

I have argued that the material of a comet consists of a solid nucleus, giving out fire and gas, enveloped in a great gaseous mass, and a tail made up of stones, possibly gradually diminishing in size as they recede from the nucleus, until the after-part of it is composed of fine dust ground from the pebbles and bowlders [sic]; while beyond this there may be a still further prolongation into gaseous matter. [The composition of comets and their tails is still a hotly debated matter in scientific circles. Donnelly's description is somewhat at variance with current accepted scientific dogma concerning the composition of comets, and it reflects the lack of information about comets which existed in his day.]

Now, we have seen that Biela's comets lost their tails. What became of them? There is no evidence to show whether they lost them in 1852, 1859, 1866, or 1872. The probabilities are that the demoralization took place before 1852, as otherwise the comets would have been seen, tails and all, in that and subsequent years. It is true that the earth came near enough in 1872 to attract some of the wandering gravel-stones toward itself, and that they fell, blazing and consuming themselves with the friction of our atmosphere, and reached the surface of our planet, if at all, as cosmic dust. But where were the rest of the assets of these bankrupt comets? They were probably scattered around in space, *disjecta membra,* floating hither and thither, in one place a stream of stones, in another a volume of gas; while the two heads had fled away . . . shorn of their splendors and reduced to first principles.

[Apparently, Donnelly felt that the gases and finer materials of the tails of the comets Biela had been torn away from the heads of the comets. These materials and gases preceded the advent of the comets, and contacted the earth in 1871. His evidence of that fact follows herewith.]

Did anything out of the usual order occur on the face of the earth about this time?

'Yes. In the year 1871, on Sunday, the 8th of October, at half past nine o'clock in the evening, events occurred which attracted the attention of the whole world, which caused the death of hundreds of human beings, and the destruction of millions of [dollars worth of] property, and which involved three different States of the Union in the wildest alarm and terror.

The summer of 1871 had been excessively dry; the moisture seemed to be evaporated out of the air; and on the Sunday above named the atmospheric conditions all through the Northwest were of the most peculiar character. The writer was living at the time in Minnesota, hundreds of miles from the scene of the disasters, and he can never forget the condition of things. There was a parched, combustible, inflammable, furnace-like feeling in the air, that was really alarming. It felt as if there were needed but a match, a spark, to cause a world-wide explosion. It was weird and unnatural. I have never seen nor felt anything like it before or since. Those who experienced it will bear me out in these statements.

At that hour, half past nine o'clock in the evening, *at apparently the same moment,* at points hundreds of miles apart, in three different States, Wisconsin, Michigan, and Illinois, fires of the most peculiar and devastating kind broke out, so far as we know, by spontaneous combustion.

In Wisconsin, on its eastern borders, in a heavily timbered country, near Lake Michigan, a region embracing *four hundred square miles,* extending north from Brown County, and containing Peshtigo, Manistee, Holland, and numerous villages on the shores of Green Bay, was swept bare by an absolute whirlwind of flame. There were *seven hundred and fifty people killed outright,* besides great numbers of the wounded, maimed, and burned, who died afterward. More than three million dollars' worth of property was destroyed. (See

"History of the Great Conflagration," Sheahan & Upton, Chicago, 1871, pp. 393, 394, etc.)

It was no ordinary fire. I quote:

"At sundown there was a lul in the wind and comparative stillness. For two hours there were no signs of danger; but at a few minutes after nine o'clock, and by a singular coincidence, *precisely the time at which the Chicago fire commenced,* the people of the village heard a terrible roar. It was that of a tornado, crushing through the forests. *Instantly the heavens were illuminated with a terrible glare. The sky,* which had been so dark a moment before, *burst into clouds of flame.* A spectator of the terrible scene says the fire did not come upon them gradually from burning trees and other objects to the windward, but the first notice they had of it was *a whirlwind of flame in great clouds from above the tops of the trees,* which fell upon and entirely enveloped everything. The poor people inhaled it, or the intensely hot air, and fell down dead. This is verified by the appearance of many of the corpses. They were found dead in the roads and open spaces, *where there were no visible marks of fire near by, with not a trace of burning upon their bodies or clothing.* At the Sugar Bush, which is an extended clearing, in some places four miles in width, corpses were found in the open road, between fences only slightly burned. *No mark of fire was upon them; they lay there as if asleep.* This phenomenon seems to explain that fact that so many were killed in compact masses. They seemed to have huddled together, in what were evidently regarded at the moment as the safest places, *far away from buildings, trees, or other inflammable* material, and there to have died together." (Ibid., p. 372.)

Another spectator says:

"Much has been said of the intense heat of the fires which destroyed Peshtigo, Menekaune, Williamsonville, etc., but all that has been said can give the stranger but a faint conception of the reality. The heat has been compared to that engendered by a flame concentrated on an object by a blowpipe; but even that would not account for some of the phenomena. For instance, we have in our possession a copper cent taken from the pocket of a dead man in the Peshtigo Sugar Bush, which will illustrate our point. *This cent has been partially fused,* but still retains its round form, and the inscription upon it is legible. Others, in the same pocket, were partially *melted,* and yet *the clothing and the body of the man were not even singed.* We do not know in what way to account for this, unless, as is asserted by some, the tornado and fire were accompanied by electrical phenomena." (Ibid., p. 373.) [It should be noted here that Velikovsky maintained that electromagnetic phenomena were an integral part of such events, although his information came from ancient sources—not recent accounts such as these.]

"It is the universal testimony that the prevailing idea among the people was, that the last day had come. Accustomed as they were to fire, nothing like this had ever been known. They could give no other interpretation to this ominous roar, this *bursting of the sky with flame, and this dropping down of fire out of the very heavens,* consuming instantly everything it touched.

"No two give a like description of the great tornado as it smote and devoured the village. It seemed as if 'the fiery fiends of hell had been loosened,' says one. 'It came in great sheeted *flames from heaven,*' says another. 'There was *a pitiless rain of fire and SAND.*' 'The atmosphere was all afire.' Some speak of '*great balls of fire unrolling and shooting forth in streams.*' The fire leaped over roofs and trees, and ignited whole streets at once. No one could stand before the blast. It was a race with death, above, behind, and before them." (Ibid., p. 374.)

A civil engineer, doing business in Peshtigo, says:

"The heat increased so rapidly, as things got well afire, that, *when about four hundred feet from the bridge and the nearest building,* I was obliged to lie down behind a log that was aground in about two feet of water, and by going under water now and then, and holding my head close to the water behind the log, I managed to breathe. There were a dozen others behind the same log. If I had succeeded in crossing the river and gone among the buildings on the other side, probably I should have been lost, as many were."

We have seen Ovid describing the people of "the earth" crouching in the same way in the water to save themselves from the flames of the Age of Fire.

In Michigan, one Allison Weaver, near Port Huron, determined to remain, to protect, if possible, some mill-property of which he had charge. He knew the fire was coming, and dug himself a shallow well or pit, made a thick plank cover to place over it, and thus prepared to bide the conflagration.

I quote:

"He filled it nearly full of water, and took care to saturate the ground around it for a distance of several rods. Going to the mill, he dragged out a four-inch plank, sawed it in two, and saw that the parts tightly covered the mouth of the little well. 'I kalkerated it would be tech and go,' said he, 'but it was the best I could do.' At midnight he had everything arranged, and the roaring then was awful to hear. The clearing was ten to twelve acres in extent, and Weaver says that, for two hours before the fire reached him, there was a constant flight across the ground of small animals. As he rested a moment from giving the house another wetting down, a horse dashed into the opening at full speed and made for the house. Weaver could see him tremble and shake with excitement and terror, and felt a pity for him. After a moment, the animal gave utterance to a snort of dismay, ran two or three times around the house, and then shot off into the woods like a rocket.

"Not long after this the fire came. Weaver stood by his well, ready for the emergency, yet curious to see the breaking-in of the flames. The roaring increased in volume, the air became oppressive, a cloud of dust and cinders came showering down, and he could see the flame through the trees. It did not run along the ground, or leap from tree to tree, but it came on like a tornado, *a sheet of flame reaching from the earth to the tops of the trees.* As it struck the clearing he jumped into his well, and closed over the planks. He could no longer see, but he could hear. He says that the flames made no halt whatever, or ceased their roaring for an instant, but he hardly got the opening closed before the house and mill were burning tinder, and both were down in five minutes. The smoke came down upon him powerfully, and his den was so hot he could hardly breathe.

He knew that the planks above him were on fire, but, remembering their thickness, he waited till the roaring of the flames had died away, and then with his head and hands turned them over and put out the fire by dashing up water with his hands. Although it was a cold night, and the water had at first chilled him, the heat gradually warmed him up until he felt quite comfortable. He remained in his den until daylight, frequently turning over the planks and putting out the fire, and then the worst had passed. The earth around was on fire in spots, house and mill were gone, leaves, brush, and logs were swept clean away as if shaved off and swept with a broom, and nothing but soot and ashes were to be seen." (Ibid., p. 390.)

In Wisconsin, at Williamson's Mills, there was a large but shallow well on the premises belonging to a Mr. Boorman. The people, when cut off by the flames and wild with terror, and thinking they would find safety in the water, leaped into this well. "The relentless fury of the flames drove them pell-mell into the pit, to struggle with each other and die—some by drowning, and others by fire and suffocation. None escaped. *Thirty-two bodies were found there.* They were in every imaginable position; but the contortions of their limbs and the agonizing expressions of their faces told the awful tale." (Ibid., 386.)

The recital of these details, horrible though they may be, seem excusable when we remember that the ancestors of our race must have endured similar horrors in that awful calamity which I have discussed in this volume.

James B. Clark, of Detroit, who was at Uniontown, Wisconsin, writes:

"The fire suddenly made a rush, like the flash of a train of gunpowder, and swept in the shape of a crescent around the settlement. It is almost impossible to conceive *the frightful rapidity of the advance of the flames.* The rushing fire seemed to eat up and annihilate the trees."

They saw a black mass coming toward them from the wall of flame:

"It was a stampede of cattle and horses thundering toward us, bellowing, moaning, and neighing as they galloped on; rushing with fearful speed, their eyeballs dilated and glaring with terror, and every motion betokening delirium of fright. Some had been badly burned, and must have plunged through a long space of flame in the desperate effort to escape. Following considerably behind came a solitary horse, panting and snorting and nearly exhausted. He was saddled and bridled, and, as we first thought, had a bag lashed to his back. As he came up we were startled at the sight of a young lad lying fallen over the animal's neck, the bridle wound around his hands, and the mane being clinched by the fingers. Little effort was needed to stop the horse, and at once release the helpless boy. He was taken into the house, and all that we could do was done; but he had inhaled the smoke, and was seemingly dying. Some time elapsed and he revived enough to speak. He told his name—Patrick Byrnes—and said: 'Father and mother and the children got into the wagon. I don't know what became of them. Everything is burned up. I am dying. Oh! is hell any worse than this?" (Ibid., p. 383.)

How vividly does all this recall the book of Job and the legends of Central America, which refer to the multitudes of the burned, maimed, and wounded lying in the caverns, moaning and crying like poor Patrick Byrnes, suffering no less in mind than in body!

When we leave Wisconsin and pass about two hundred and fifty miles eastward, over Lake Michigan and across the whole width of the State of Michigan, we find much the same condition of things, but not so terrible in the loss of human life. Fully *fifteen thousand people were rendered homeless by the fires;* and their food, clothing, crops, horses, and cattle were destroyed. Of these five to six thousand were burned out *the same night that the fires broke out in Chicago and Wisconsin.* The total destruction of property exceeded one million dollars; not only villages and cities, but whole townships, were swept bare.

But it is to Chicago we must turn for the most extraordinary results of this atmospheric disturbance. It is needless to tell the story in detail. The world knows it by heart:

> "Blackened and bleeding, helpless, panting, prone,
> On the charred fragments of her shattered throne,
> Lies she who stood but yesterday alone."

I have only space to refer to one or two points.

The fire was spontaneous. The story of Mrs. O'Leary's cow having started the conflagration by kicking over a lantern was proved to be false. It was the access of gas from the tail of Biela's comet that burned up Chicago!

The fire-marshal testified:

"I felt it in my bones that we were going to have a burn."

He says, speaking of O'Leary's barn:

"We got the fire under control, and it would not have gone a foot farther; but the next thing I knew they came and told me that St. Paul's church, *about two squares north, was on fire.*" (Ibid., p. 163.)

The checked the church-fire, but—

"The next thing I knew the fire was in Bateham's planing-mill."

A writer in the New York *Evening Post* says he saw in Chicago "buildings far beyond the line of fire, *and in no contact with it,* burst into flames from the interior."

It must not be forgotten that the fall of 1871 was marked by extraordinarry conflagrations in regions widely separated. On the 8th of October, *the same day* the Wisconsin, Michigan, and Chicago fires broke out, the States of Iowa, Minnesota, Indiana, and Illinois were severely devastated by prairie-fires; while terrible fires raged on the Alleghenies, the Sierras of the Pacific coast, and the Rocky Mountains, and in the region of the Red River of the North.

The Annual Record of Science and Industry for 1876, page 84, says:

"For weeks before and after the great fire in Chicago in 1872, great areas of forest and prairie-land, both in the United States and the British Provinces, were on fire."

The flames that consumed a great part of Chicago were of an unusual character and produced extraordinary effects.

They absolutely *melted* the hardest buildingstone, which had previously been considered fire-proof. Iron, glass, granite, were fused and run together into grotesque conglomerates, as if they had been put through a blast-furnace. No kind of material could stand its breath for a moment.

I quote again from Sheahan & Upton's work:

"The huge stone and brick structures melted before the fierceness of the flames as a snow-flake melts and disappears in water, and almost as quickly. Six-story buildings would take fire and *disappear for ever from sight in five minutes by the watch* . . . . The fire also doubled on its track at the great Union Depot and burned half a mile southward *in the very teeth of the gale*—a gale which blew a perfect tornado, and in which no vessel could have lived on the lake . . . . *Strange, fantastic fires of blue, red, and green played along the cornices of buildings.*" (*History of the Chicago Fire*, pp. 85, 86.)

Hon. William B. Ogden wrote at the time:

"The fire was accompanied by the fiercest tornado of wind ever known to blow here." (Ibid., p. 87.)

"The most striking peculiarity of the fire was its intense heat. Nothing exposed to it escaped. Amid the hundreds of acres left bare there is not to be found a piece of wood of any description, and, *unlike most fires, it left nothing half burned.* . . . The fire swept the streets of all the ordinary dust and rubbish, consuming it instantly." (Ibid., p. 119.)

The Athens marble burned like coal!

"The intensity of the heat may be judged, and the thorough combustion of everything wooden may be understood, when we state that in the yard of one of the large agricultural-implement factories was stacked some hundreds of tons of pig-iron. This iron was two hundred feet from any building. To the south of it was the river, one hundred and fifty feet wide. No large building but the factory was in the immediate vicinity of the fire. Yet, so great was the heat, that *this pile of iron melted and run, and is now in one large and nearly solid mass.*"(Ibid., p. 121.)

The amount of property destroyed was estimated by Mayor Medill at one hundred and fifty million dollars; and the number of people rendered houseless, at one hundred and twenty-five thousand. Several hundred lives were lost.

All this brings before our eyes vividly the condition of things when the comet struck the earth; when conflagrations spread over wide areas; when human beings were consumed by the million; when their works were obliterated, and the the remnants of the multitude fled before the rushing flames, filled with unutterable consternation . . . .

Who shall say how often the characteristics of our atmosphere have been affected by accessions from extraterrestrial sources, resulting in conflagrations or pestilences, in failures of crops, and in famines? Who shall say how far great revolutions and wars and other perturbations of humanity have been due to similar modifications?

Appendix B

Over the years, many references to the following vision have been made in the writings of L.D.S. authors, though it has not been generally available to the Saints.

In order that the student of prophecy might have the opportunity to read it for himself, it is reprinted here. Though this revelation is usually attributed to Wilford Woodruff because it appears in his journal, the author of the book from which this account was taken maintains that the vision was given to John Taylor. His evidence for that rests in the fact that the narrator of the vision speaks of "reading the revelations in the French Language." John Taylor alone, among the presiding authorities in that day, spoke French. Wilford Woodruff did not speak French, let alone read it.

- - - - - - - - - - -

A Vision [apparently] given to President John Taylor at Salt Lake City, on the evening of December 16, 1877.

I went to bed at my usual hour half past nine o'clock. I had been reading the revelations in the French language. My mind was calm, more so than usual if possible to be so. I composed myself for sleep but could not sleep. I felt a strange stupor come over me and apparently became partially unconscious, still I was not asleep, nor awake with a strange far away dreamy feeling.

The first thing I recognized was that I was in the Tabernacle at Ogden sitting in the back seat in the corner for fear they would call upon me to preach, which, after singing the second time, they did, by calling me to the stand. I arose to speak and said I did not know that I had anything special to say except to bear my testimony to the truth of the latter-day work. When all at once it seemed as though I was lifted out of myself, and I said "Yes, I have something to say, it is this: some of my Brethren present have been asking me what is coming to pass, what is the wind blowing up. I will answer you right here what is coming to pass shortly."

I was immediately in Salt Lake City wandering about the streets in all parts of the city, and on the door of every house I found a badge of mourning, and I could not find a house but what was in mourning. I passed by my own house and saw the same sign there and asked, "Is that me that is dead?" Something gave me an answer, "No, you'll live through it all."

It seemed strange to me that I saw no person on the street in my wandering about through the city. They seemed to be in their houses with their sick and dead. I saw no funeral procession, or anything of the kind, but the city looked very still and quiet as though the people were praying, and had control of the disease whatever it was. I then looked in all directions over the territory, east, west, north, and south and I found the same mourning in every place throughout the land.

The next I knew I was just this side of Omaha. It seemed as though I was above the earth, looking down to it as I passed along on my way east. I saw the roads full of people, principally women, with just what they could carry in bundles on their backs, traveling to the mountains on foot, and I wondered how they could get there with nothing but a small pack upon their backs. It was remarkable to me that there were so few men among them. It did not seem as though the cars were running. The rails looked rusty and the road abandoned and I have no conception how I traveled myself.

As I looked down upon the people I continued eastward through Omaha and Council Bluffs which were full of disease and women everywhere. The states of Missouri and Illinois were in turmoil and strife, men killing each other, and women joining in the fight, family against family, cutting each other to pieces in the most horrid manner.

The next I saw was Washington and I found the city a desolation. The White House empty, the halls of Congress the same, everything in ruins, the people seemed to have fled from the city and left it to take care of itself.

I was next in the city of Baltimore, and in the square where the monument of 1812 stands in front of St. Charles. And at the hotels, I saw the dead piled up so as to fill the square. I saw mothers cut the throats of their own children for the sake of their blood, which they drank from their veins to quench their thirst and then lie down and die. The waters of the Chesapeake [River] and of the city were so stagnant and such a stench arose from them on account of the putrification of dead bodies that the very smell caused death.

And what was singular again, I saw no men except they were dead, lying in the streets, and very few women, and they were crazy mad, and in a dying condition. Everywhere I went I beheld the same all over the city, and it was horrible beyond description to look at.

I thought this must be the end. But no, I was seemingly in

Philadelphia and there [as in Baltimore] everything as still. No living soul was to be seen to greet me, and it seemed as though the whole city was without an inhabitant. In Arch and Chestnut Street[s] and in fact everywhere I went, the putrification of the dead bodies caused such a stench that it was impossible for any creature to exist alive, nor did I see any living thing in the city.

I next found myself in Broadway, New York, and there it seemed the people had done their best to overcome the disease. But in wandering down Broadway I saw the bodies of beautiful women lying stone dead, and others in a dying condition on the sidewalk. I saw men crawl out of the cellars and rob the dead bodies of the valuables they had on [them], and before they could return to their coverts in the cellars, they themselves would roll over a time or two and die in agony. On some of the back streets I saw mothers kill their own children and eat raw flesh and then in a few minutes die themselves.

Wherever I went I saw the same scenes of horror and desolation, rapine and death. No horses or carriages, no buses or streetcars, but death and destruction everywhere.

I then went to the Grand Central Park, and looking back, I saw a fire start and just at that moment a mighty east wind sprang up and carried the flames west over the city, and it burned until there was not a single building left standing whole, even down to the wharfs, and the shipping all seemed to be burned, and swallowed up in the common destruction and left nothing but a desolation where the great city was, a short time before. The stench from the bodies that were burning was so great that it was carried a great distance across the Hudson River and Bay, and thus spread disease and death wherever the flames penetrated.

I cannot paint in words the horror that seemed to

encompass me around. It was beyond description or thought of man to conceive. I supposed this was the end, but I was here given to understand that the same horror was being enacted all over the country, north, south, east and west, that few were left alive, still there were some.

Immediately after, I seemed to be standing on the west bank of the Missouri River opposite the City of Independence, but I saw no city. I saw the whole states of Missouri and Illinois and part of Iowa were a complete wilderness with no living human being in them.

I then saw a short distance from the river twelve men dressed in the robes of the Temple stand in a square or nearly so. I understood it represented the twelve gates of the New Jerusalem, and they were, with hands uplifted, consecrating the ground and laying the cornerstones. I saw myriads of angels hovering over them and around about them and also an immense pillar of a cloud hover over them. And I heard the singing of the most beautiful music. The words [were] "Now is established the Kingdom of our God and His Christ and He shall reign forever and ever, and the Kingdom shall never be thrown down, for the Saints have overcome."

And I saw people coming from the river and [from] different places a long way off to help build the Temple, and it seemed that the hosts of the angels also helped to get the material to build the Temple. And I saw some come who wore their Temple robes to help build the Temple and the city, and all the time I saw the great pillar of cloud hovering over the place.

Instantly I found I was in the Tabernacle at Ogden [and] yet I could see the building going on, and [I] got quite animated in calling to the people in the Tabernacle to listen to the beautiful music that the angels were making. I called to them to look at the angels as the house seemed to be full of them and they were saying the same words that I heard before: "Now is the Kingdom of our God [and His Christ] established forever and ever."

And then a voice said, "Now shall come to pass that which was spoken by Isaiah the Prophet, 'That seven women shall take hold of one man saying, &c.'"

At this time I seemed to stagger back from the pulpit and F. D. Richards and someone else caught me and prevented me from falling, when I requested Brother Richards to apologize to the audience for me because I stopped so abruptly, and tell them I had not fainted but was exhausted.

I [then] rolled over in my bed and heard the City Hall clock strike twelve [o'clock].

(*Wilford Woodruff Journals,* June 15, 1878, Church Historians Office.)

. . .And The Earth Shall Reel To And Fro. . .

Bibliography

Abnormal Psychology. E. Rosen and I. Gregory. Philadelphia and London, 1965.

Annual Record of Science and Industry, 1876.

Aurelius Augustinus, *The City of God* (De Civitates Dei), XXI, 8 transl. by G. Walsh and G. Monahan (New York, 1952), vol. III.

Brown, P. L. *Comets, Meteorites & Men,* New York, 1973, pp. 10-14.

The Contributor. Published by the Church of Jesus Christ of Latter-day Saints, August 1894, Salt Lake City, Utah, Vol. 15, No. 20, pp. 638-647.

Crowther, Duane S. *Prophecy—Key To The Future,* Bookcraft, Salt Lake City, Utah, 1962.

Darwin, Charles. *On the Origin of Species by Means of Natural Selection.*

Desmond, Adrian J. *The Hot-Blooded Dinosaurs* (New York, 1976).

Documentary History of the Church. Deseret Book Co., Salt Lake City, Utah, 1978.

Donnelly, Ignatius. *Ragnarok: The Age of Fire and Gravel,* New York, D. Appleton and Company, 1883.

Gardiner, A. H. *Admonitions of an Egyptian Sage* from a hieratic papyrus in Leiden, 1909.

History of the Great Conflagration. Sheahan & Upton, Chicago, Illinois, 1871.

Inspired Prophetic Warnings. Nels B. Lundwall, Salt Lake City, Utah, 1940.

James, E. O. *The Ancient Gods,* New York, G. P. Putnam's Sons, 1960.

Journal of Discourses.

Journals of Wilford Woodruff, B.Y.U. Special Collections.

Kronos, A Journal of Interdisciplinary Synthesis. Glassboro, New Jersey, Kronos Press.

Lindsey, Hal and Carlson, C. C. *The Late Great Planet Earth,* Grand Rapids, Michigan: Zondervan Publishing House, 1970.

Millennial Star.

Ovid. *The Metamorphoses,* Book II, Phaethon's Ride, translated by Horace Gregory, New York: Mentor Books, 1960.

Patten, Donald W. *The Biblical Flood and the Ice Epoch,* Seattle, Washington: Pacific Meridian Publishing Company, 1966.

Pensee. The magazine for the Student Academic Freedom Forum, published by David N. Talbott, Steven L. Talbott, ed. Portland: Oregon.

Plato. *Timaeus,* translated by F. M. Cornford, London: Routledge & Kegan Paul, 1937.

Ransom, C. J. *The Age of Velikovsky,* Glassboro, New Jersey: Kronos Press, 1976.

Saturday Night Thoughts. Deseret News, Salt Lake City, Utah, 1921, Orson F. Whitney.

Skousen, W. Cleon. *Prophecy and Modern Times,* Salt Lake City, Utah, Deseret News Press, 1939.

Smith, Joseph Fielding. *Teachings of the Prophet Joseph Smith,* Deseret Book, Salt Lake City, Utah, 1963, c. 1938.

Smith, Robert W. *The Last Days.* Published by Robert W. Smith, Pyramid Press, Salt Lake City, Utah, 1947.

Smith, William. *The New Smith's Bible Dictionary,* Garden City, New York: Doubleday & Company, 1979.

Talbott, Stephen L., ed. *Velikovsky Reconsidered,* New York, New York: Warner Books, 1977.

The Book of Jasher, Salt Lake City, Utah: J. H. Parry & Company, 1887.

Velikovsky and Establishment Science, Kronos Press: Glassboro, New Jersey, 1977.

Velikovsky, Immanuel, *Ages in Chaos,* Garden City, New York: Doubleday & Company, 1952.

Velikovsky, Immanuel. *Earth in Upheaval,* Garden City, New York: Doubleday & Company, 1955.

Velikovsky, Immanuel. *Theses for the Reconstruction of Ancient History,* 1945.

Velikovsky, Immanuel. *Worlds in Collision,* Garden City, New York: Doubleday & Company, 1950.